'Real Girls Don't Do Maths'
Gender and the Construction of Privilege

Sue Willis
Murdoch University

Deakin University

This book forms part of ECS807 *Teaching and Learning Mathematics*, which is one of the units offered by the School of Education in Deakin University's Open Campus Program. It has been prepared for the Teaching and Learning Mathematics team.

The study materials include:
M. A. (Ken) Clements, *Mathematics for the Minority: Some Historical Perspectives of School Mathematics in Victoria*
N. F. Ellerton & M. A. (Ken) Clements (eds), *School Mathematics: The Challenge to Change*
P. Harris, *Mathematics in a Cultural Context: Aboriginal Perspectives on Space, Time and Money*
N. A. Pateman, *Teaching Mathematics—a Tantalising Enterprise: On the Nature of Mathematics and Mathematics Teaching*
S. Willis, *'Real Girls Don't Do Maths': Gender and the Construction of Privilege*

More titles may be added to this list from time to time.
These books are available from Deakin University Press, Deakin University, Geelong, Victoria 3217. Enrolled students also receive a guide to the unit.

Published by Deakin University, Geelong, Victoria 3217
Distributed by Deakin University Press
First published 1989
Reprinted 1991

© Deakin University 1989

Edited, designed and typeset by Deakin University Book Production Unit
Printed by Deakin University Printery

National Library of Australia
Cataloguing-in-publication data

Willis, Sue 1948– .
 'Real girls don't do maths'.

 Bibliography.
 ISBN 0 7300 0631 X.

 1. Mathematics—Study and teaching. 2. Sex differences in education. 3. Mathematical ability—Sex differences. I. Deakin University. School of Education. Open Campus Program. II. Title.

501'.7

Contents

Chapter 1 Girls and mathematics	1
Introduction	1
An overview of the changing definition of the 'problem of girls and mathematics'	2
Chapter 2 The evidence of gender differences in achievement and participation in mathematics	4
The evidence of achievement in the United States of America prior to 1974	4
Recent Australian evidence of mathematics achievement in Years K–10	5
Overseas evidence of mathematics achievement in Years K–10	8
Recent Australian evidence of mathematics achievement in Years 11–12	9
The evidence of gender differences in the extremes of mathematics achievement	10
The evidence of gender differences in participation in mathematics	11
Questionable uses of the available evidence	13
Chapter 3 Explanations of gender differences in giftedness in mathematics	16
Some cross-cultural evidence	16
Opportunities to learn mathematics	18
Confidence, anxiety and success in competitions	20
The nature of the test and the testing situation	24
Chapter 4 Explanations of gender differences in participation in mathematics	26
Confidence, anxiety and participation	26
Appropriateness of the choice of mathematics	28
Opinion of the subject mathematics	31
Chapter 5 Which way forward?	34
Why more mathematics?	34
Towards an empowering mathematics curriculum	37
Towards a mathematics curriculum inclusive of girls	38
Conclusion	39
References	40
Annotated bibliography	47

Chapter 1
Girls and mathematics

Introduction

In 1987, as part of a federal election campaign, the Prime Minister of Australia promised that his government, if re-elected, would set up a national program to increase the participation of girls in mathematics and science. It would be easy to be cynical about election promises generally and this one in particular. None the less, this promise was interesting in several ways, not the least being that it was presented as an *election* promise and, presumably, expected to bring the government more votes than it would lose. Twenty years ago, such a promise would have been widely regarded as outlandish and as evidence of a government which, at best, was pandering to the 'women's libbers' and, at worst, had lost all track of reality. But what we regard as reality has changed and, in 1987, such a promise, whether in fact a 'vote getter' or not, was presented as part of a rational and responsible economic plan and was acceptable in the political mainstream of Australia.

The purpose of this monograph is to discuss issues of gender and mathematics education. As evidenced by the election promise, a particularly highly profile issue at the moment is that girls participate less than boys in mathematics. For some, the issue is one of social justice, since girls do not share equally in the advantages of the mathematically well prepared. For others, the issue is the national need to advance technologically; girls do not provide, they argue, a sufficiently high mathematical 'yield' to the nation. Regardless of the perspective taken, recent years have seen a growing acceptance, in educational circles but more widely also, of the notion that there is a 'problem of girls and mathematics'. Reaching agreement about how to interpret this problem and what we would regard as a solution are, however, proving both elusive and contentious.

I begin this monograph by briefly overviewing the way in which our interpretation of the 'problem of girls and mathematics' has changed over the past several decades. Following this, I review the evidence of gender differences in achievement and participation in mathematics, describe some of the major explanations offered for these differences and then consider the implications for the practice of mathematics education. In so doing, I present some alternative conceptions of the 'problem', its nature and solutions, and argue that the lower participation in mathematics of girls

I should like to thank Margaret Clark, Margo Haysom and Jane Kenway for their invaluable feedback on earlier drafts of this monograph.

than of boys reflects a problem not only for girls but also for boys and for the community generally. It is not a problem of or with individual girls or girls collectively but, rather, it is a problem of and with mathematics education, the way it is defined and the uses to which it is put.

In this monograph I do not address school- and system-level policy decisions which influence both the educational experiences of students and their post-school options. Certainly, policy implications could be drawn from the state participation data that is included and other implications could be drawn from an analysis of the difference between preparatory and selective uses of the mathematics curriculum in primary and secondary schools. I choose instead to focus on classroom practice in mathematics because it is in classrooms that the long-term and fundamental educational solutions lie. This is not an issue that can be left to the committed, the experts or the policy makers — it must be addressed by every teacher of mathematics.

An overview of the changing definition of the 'problem of girls and mathematics'

Over one hundred years ago there were arguments about the inclusion of mathematics in the curriculum for girls. Even amongst those most committed to girls learning mathematics, however, many believed that girls were physically, emotionally and intellectually ill-equipped for the serious study of mathematics. For example, Dorothea Beale, a leader in the higher education for girls movement, stated in 1868:

> ... I do not think that the mathematical powers of women enable them generally — (their physical strength, I dare say, has a great deal to do with it) to go so far in the higher branches, and I think we should be straining the mind (which is the thing of all things to be most deprecated) if we were to try to force them to take up such examinations ... (quoted in Clements 1979, p. 317)

In Australia twenty years ago, it was still widely and complacently believed that girls are mathematically less able than boys. While most of us had grown past the view expressed by Dorothea Beale, on the whole we regarded the existence of gender differences in mathematics achievement unproblematically since it provided yet another confirming instance of our deep-rooted belief that women were the weaker sex — physically and intellectually. Indeed, the commonsense notion of the 1960s was that girls could not, need not and would not want to do advanced mathematics — or at least not to the same extent as boys.

During the 1960s researchers began to 'explain' girls' low achievement in mathematics. ('Why *can't* girls do as well as boys in mathematics?' they asked.) In 1964, Smith, a Scottish psychologist, published a book in which he argued that sex differences in mathematical *ability* could be explained by genetically determined differences in spatial ability. At the time of publication, the book and the issue were of little interest outside the particular academic field of psychology. Within a decade, however, the climate had changed both in Australia and internationally. In educational circles most people now talked of mathematical *achievement* rather than of mathematical ability (although talk of the male mathematical gene continues to appear in the media with disconcerting regularity). Mathematical and spatial achievement were at the centre of a nature/nurture debate on gender differences, fuelled in part by competing psychological theories of intelligence, but also by the various political movements of the 1960s and 1970s which exposed as flawed the biological determinism used to justify class, race and gender differences in educational and occupational experiences.

By the early 1970s, the reputed lower achievement and participation of girls in mathematics was regarded as a problem — especially for the girls! Many people had come to consider it appropriate that girls be prepared for a wider range of employment opportunities than they were previously. It was still understood, of course, that girls achieve less well than boys in mathematics, but there had been a subtle, although significant, shift; less often did we ask, 'Why *can't* girls do as well as boys in mathematics?' and more often, 'Why *don't* girls do as well as boys in mathematics?'

A considerable research literature which focused upon 'affective influences' on girls' achievement in mathematics developed during the 1970s. For primary and secondary school levels, and contrary to some common beliefs, little evidence was found to show that mathematics was enjoyed by girls any less than by boys, and attempts to relate mathematics achievement to liking or disliking the subject, on the whole, failed (see reviews by Atweh 1980 and Schofield 1982). Similarly, explanations of difference were sought in studies of 'achievement motivation' which were, in turn, related to differential societal expectations of girls and boys with regard to assertiveness, independence, competitiveness, 'success avoidance' and so on. In Australia, Leder (1977, 1980a, 1982), following Horner (1972), studied the relevance of the 'fear of success' construct to 'mathematically able' girls. The effect of task variables, such as the context in which problems are set, on the relative success of girls and boys (Leder 1974, 1976) and the effect of the sex stereotyping of textbooks (Carss 1982; Grimison 1982) were also explored. The belief that girls do not do well at mathematics remained largely unchallenged until the late 1970s when researchers began to express doubts even about that.

It is perhaps only a slight oversimplification to suggest that the commonsense conception of girls and mathematics thirty years ago was that there was no problem, it was 'only natural' that girls could not, did not and would not want to do mathematics. Twenty years ago, however, that girls *could not* do as well as boys in mathematics had come to be regarded as a problem worthy of the attention of psychologists and educators. Ten years later, fuelled in part by the increasing popularity of 'social' explanations of educational achievement, but also by the re-emergence of feminism amidst social, political and economic change, the problem had also changed. We no longer asked why girls could not but rather why they *did not* do as well as boys at mathematics. Now, in the late 1980s, the problem has become why girls *will not* do as well as boys, why they choose to participate less. For many who are concerned with equal opportunity, today's problem is that girls limit their post-school options by choosing not to study certain mathematics courses; for others, it is the waste of talent implied when girls, as a group, 'underparticipate' in mathematics. In either case, the solution is seen to be to require that more girls be encouraged to undertake more mathematics.

Chapter 2
The evidence of gender differences in achievement and participation in mathematics

Much of the early interest in gender differences in mathematics began in the United States of America. Consequently, this chapter begins with a brief look at an extremely influential North American review of the evidence of gender differences. Following this, the Australian evidence is reviewed, and similarities to and differences from overseas evidence are highlighted. Finally, attention is drawn to some of the more questionable interpretations and uses of the evidence of gender differences in mathematics achievement and participation.

The evidence of achievement in the United States of America prior to 1974

In 1974, Maccoby & Jacklin carried out an extensive review of North American research related to gender differences and published what became regarded as the definitive statement on the matter. They concluded that the majority of beliefs about intellectual and emotional differences between males and females were unfounded but that three intellectual differences were fairly well established. These were that females have greater verbal abilities than males, while males have greater visual–spatial abilities and greater mathematical abilities than females.

Later analyses of the studies reviewed by Maccoby & Jacklin cast some doubt on the educational relevance of their conclusions. Firstly, of the total of 160 studies of verbal ability that were reviewed, 100 (62.5%) showed no difference between boys and girls, 45 (28%) showed that girls were superior and 15 (9%) showed that boys were superior. Similarly, of 38 comparisons of quantitative ability, there was no difference in 18 comparisons (47%), in 4 comparisons (11%) girls were superior and in 16 (42%) boys were superior. The trends were in the direction indicated by Maccoby & Jacklin but the results were more equivocal than was often assumed. Meta-analyses of the same studies suggest that, although at the time of the studies gender *was* an influence on verbal and mathematical achievement, it accounted for only 1% of the variance in each of verbal and quantitative achievement, and the mean achievement levels were less than one-quarter of a standard deviation apart for verbal ability and less than one-half for quantitative ability (Hyde 1981; Rosenthal & Rubin 1982). The most recent study reviewed by Maccoby & Jacklin is now more than 15 years old and all were North American. Even at the time, differences in verbal and quantitative ability were educationally insignificant for the majority of boys and girls, yet this study is still quoted, even in Australia, as though it were definitive and relevant (see, for example, Yates & Firkin 1986, p.2; Day 1987, p. 9). Also, most studies prior to 1974 failed to control for differential course-taking and school-leaving patterns. There is some reason to believe that the least mathematically able boys are likely to be the earliest school leavers but even without this, given the widely held belief that girls could not do mathematics and given the optional nature of the serious study of mathematics in the United States of America, courses taken by the boys and girls often differed considerably.

What is interesting is that so many of us found it very easy to accept uncritically the existence of differences between boys' and girls' achievement in language and mathematics and were prone to exaggerate both their extent and their educational and occupational relevance. Some of the dangers of this exaggeration are self-evident, others will become clearer as we proceed. Meanwhile, what is the current Australian evidence?

Recent Australian evidence of mathematics achievement in Years K–10

Australian research suggests that, in the primary and lower secondary years, overall levels of achievement of boys and girls are very similar, all other things being equal. Australia participated in the first (1964) and second (1978) International Educational Assessment (IEA) Mathematics Studies. For these studies, multiple choice tests were designed co-operatively by the countries involved and were administered to students in schools chosen by stratified random sampling. Regression analyses of the Australian data showed that, for all states except Western Australia in 1964 and Victoria in 1978, at 13 years of age, 'sex of the student was not significantly associated with . . . the criterion of mathematics achievement' (Rosier 1980, p. 134). While, in 1978, 13-year-old boys outperformed girls of the same age in Victoria, the existence of nearby technical schools may have changed the population of some of the schools that were tested. At the time, most such technical schools were for males only, and the males remaining in regular schools were typically a more select group. This distinction between technical and regular schools was peculiar to Victoria, and the exact influence of this on Victorian data on gender differences is not clear.

A more recent study (Parker & Offer 1987) analysed Year 10 achievement in Western Australia in the period from 1972 to 1986 during which the structure of lower secondary school subject choice was based on a system called the 'Achievement Certificate'. Under this system, students could not choose the level of mathematics to be taken but, instead, studied at one of four levels of mathematics based on their previous achievement. Students were allocated to levels at the beginning of each of Years 8–10 but sometimes also each term. Students' final awards of pass or credit pass at one of the levels were based on internal school assessments and were intended to reflect their final achievement level, not simply the level at which they studied. On the basis of state-wide tests (early in Years 8 and 10), the Board of Secondary Education (BSE) allocated to each school a certain number of places at each of the four levels in such a way that in the state as a whole 25% of students received an Advanced award, 25% an Ordinary award, 25% an Elementary award, 23% a Basic award and 2% no award. Schools which awarded grades inconsistent with the proportions indicated by the BSE were required to justify doing so. Two principles underpinning this structure (which changed in 1988) were that the core curriculum a student received was dependent on previous achievement and not on subject choice, and that comparability of grades awarded between schools was maintained.

For the first four years of the Achievement Certificate (1972-1975) more boys than girls received Advanced awards in mathematics. Since 1978, however, more girls than boys have done so and, indeed, during this period girls have also received more Ordinary awards. Boys have received the majority of Basic awards and more boys than girls received no award. (Note that in Western Australia almost all children stay at secondary school until the end of Year 10.) Without a more detailed analysis of individual achievement, one cannot infer that boys achieved significantly less well than girls, but one can infer that girls achieved at least as well and possibly better. Interestingly, Parker & Offer (1987) comment that the most frequent initial reaction of educators confronted with the data is surprise. As I will argue later, we tend to expect superior achievement in mathematics from boys, and when presented with evidence to the contrary we seem driven to find a way to 'explain it away'.

In Australia and internationally, gender differences reliably appear in certain aspects of mathematics even during the primary school years. In 1977, Clements & Wattanawaha reported a study of the mathematics achievement (as measured by the Monash Assessment of Mathematical Performance test of short answer items) of almost 4000 children of 10-14 years of age in 40 Victorian schools. They found that, at each year level, the girls outperformed the boys on the logic and number subtests and, at three of the four year levels, the boys outperformed the girls on the space subtest. Overall, the boys and girls did equally well on 14 items (3 logic, 5 space, 6 numeration), the girls did better on 16 items (5 logic, 1 space, 10 numeration) and the boys did better on 3 items (0 logic, 2 space, 1 numeration). The authors reviewed earlier Australian studies and found these to be generally consistent with the results provided here.

If such data reflect real differences in the aspects of the mathematics curriculum in which girls and boys excel, for whatever reason, then it is clear that the direction and extent of gender differences on any particular test will depend on the items in the test. In 1976, Keeves & Bourke, on the basis of Australian Council of Educational Research (ACER) test data collected on about 13 000 students across Australia, found that at 10 years of age girls outperformed boys on what they called numeration but that at 14 years of age girls outperformed boys on 13 items, boys outperformed girls on 16 items and for 4 items there were no differences. As Clements & Wattanawaha (1977) commented, the ACER numeration test included items on measurement and estimation, and there is evidence that items of these types are more likely to favour boys (see, for example, Pattison & Grieve 1984; Hanna & Kuendiger 1986).

This raises a point regarding the labels given to various tasks set to students. It is often stated that girls are more successful on 'computation' and boys on 'problem solving' (Armstrong 1980; Shuard 1986) or mathematical 'reasoning' (Benbow & Stanley 1980), and the inference is made, explicitly (Moss 1982) or implicitly, that boys are better at higher level tasks. It is not clear, however, that this is a valid interpretation. To explain why will necessitate a small detour.

Research on mathematics achievement is almost invariably content oriented and based on tasks which are easily assessed in written examinations and, often, in short answer or multiple choice modes. Some of the newer curriculum developments, however, focus on the development of mathematical processes through the use of investigation work. It is often argued that such investigations relate much more closely to the act of mathematics than does more traditional school mathematics. Mathematical processes include such behaviours as 'explain contradictions', 'correctly use transitive inference', 'recognize and expose a circular argument', 'select and use appropriate data in context' and 'assess the relevance of data' (Galbraith 1986). Certainly, few regard such mathematical processes as requiring 'low level' or algorithmic thinking. What may such developments mean for gender differences in mathematics achievement? In 1986, Galbraith reported a study of the use of mathematical strategies or processes by 334 students in Years 8, 9 and 10 in two Queensland high schools. The students, who were studying a traditional mathematics curriculum with no particular emphasis on mathematical processes, were given a test of structured questions designed to assess their ability to use mathematical processes. Interestingly, in each year group the girls outperformed the boys. The ACER operations test which assesses performance on such tasks as, for example, $(900 \div 30) \div 10 = \Delta \div (30 \div 10)$, was also given to the students, and the girls outperformed the boys on this test too. The superior performance on the test of mathematics processes remained significant even after statistically controlling for the effect of the abilites reflected by the operations test. In another Australian study, Bourke & Stacey (1988) tested primary school age students and found that girls also do at least as well as and at times better than boys on unfamiliar problem-solving tasks.

To return, then, to my earlier point, such evidence suggests that we should not be too ready to assume that the mathematical tasks in which girls and boys, respectively, excel can be explained in terms of cognitive level. An inspection of the IEA items described by Rosier (1980) and Moss (1982) as lower and higher level (based on Bloom's Taxonomy) indicates that many of the tasks described as higher level arithmetic or algebra had a strong spatial or measurement component, whereas the lower level arithmetic and algebra tasks were more likely to be decontextualised. Furthermore, it was unclear what criteria were used to distinguish, for example, knowledge from translation or comprehension, since these classifications depend to a considerable extent on previous learning. What may be a high-level problem-solving task to one student may require little more than rule following for another. Similarly, many of the so-called 'problem-solving' tasks on North American tests of mathematical achievement are exercises written 'in words', which many students will have been taught to deal with algorithmically. Although different from decontextualised computation, such tasks are often not more difficult, let alone of a higher level. Indeed, there is a developing consensus that tasks which are embedded in a context which has meaning for the student may often be easier because the student is able to bring 'human sense' (Donaldson 1978, p. 17) to bear on them. Good problem solvers work as much with the content of a problem as with the mathematical form, and there is considerable evidence that, in subtle and not-so-subtle ways, the content of many mathematical tasks favours boys' interests and experiences (Leder 1974; Barnes 1987/8). As Chipman (1981) suggests, familiarity of content can make a difference, and girls do well on those areas (computation, logic, algebra) where content bias is least likely. We must be careful that we understand precisely what is being measured when we assess 'mathematics' and should ensure that we are not simply interpreting evidence in ways that confirm our existing beliefs. We could interpret the pattern of achievement presented so far as indicating that girls are better at abstract tasks (disembedded) and boys at context-bounded (embedded) tasks, but we have not. My point is not that we should interpret the data in this way, but rather to suggest that we think carefully about our chosen interpretation and ask whether it reflects the evidence or only our habits of thought on the matter. Why *do* we hear quite frequently of boys' success at 'problem solving' but quite rarely of girls' success at 'logic' (or, for that matter, more recently, with mathematical investigations)?

Overseas evidence of mathematics achievement in Years K–10

A further perspective of this issue is offered by studies from the United Kingdom. For example, Shuard (1986) reported the results of a Schools Council study between 1972 and 1975 of the mathematical achievement of a national sample of 2300 10-year-old children which found that girls did better on the items ranked by teachers as most important. Furthermore, the items on which girls achieved best were those which were statistically found to be easiest for the group as a whole. A reasonable inference from this data is that differences between girls' and boys' performance are least on tasks emphasised more in class and greatest on tasks not emphasised in class. Indeed, North American researchers are increasingly concluding that when tests cover only material which is almost exclusively taught and learned in the classrooms from which the students come, no gender differences tend to appear (for example, Senk & Usiskin 1983; Rallis & Ahern 1986; Smith & Walker 1988). It may be that girls do better on more familiar tasks and boys on less familiar tasks. Equally, it may be that experiences gained outside mathematics classrooms make some apparently 'unfamiliar' tasks relatively more familiar to boys than to girls. I will return to this matter later.

It is often assumed that the results of overseas studies can be applied, almost uncritically, to Australian contexts. In particular, evidence of gender differences in North America and the United Kingdom are often quoted with no indication that the studies are not Australian, as though the origin of such research is irrelevant. This would be the case only if all differences in mathematical achievement were genetic. Clearly, there are considerable social and educational similarities among Australia, North America and the United Kingdom but there are differences also and, as international surveys show, patterns and levels of achievement differ among these countries. While an analysis of the 1964 IEA Mathematics Study (Husen 1967) showed that in most countries boys aged 13–14 years outperformed girls of the same age, there was a significant interaction between gender and country effects. In 1986, Hanna & Kuendiger published an analysis of the corresponding 1978 IEA Mathematics Study achievement data from 20 countries and found that at 13–14 years of age girls were more successful than boys in Belgium, Thailand, Finland and Hungary, while boys were more successful than girls in France, Nigeria, Israel and the Netherlands. International results on the subtests were as follows: on arithmetic, boys and girls achieved equally overall (zero total difference) but the differences by item varied from -9 (boys more successful) to $+8$ (girls more successful); on algebra, boys were slightly less successful than girls overall with differences by item of -3 to $+4$; on statistics, boys again were slightly less successful than girls overall with differences by item of -4 to $+3$; on measurement, boys were slightly more successful than girls overall with differences by item of -9 to $+5$; and on geometry, boys were considerably more successful than girls overall with differences by item of -7 to $+6$ but with an outlying score of -22! Clearly, a great many factors can contribute to achievement. In particular, there are likely to be substantial differences in children's formal and informal opportunities to learn certain mathematical skills and concepts, and the extent to which these occur broadly along gender lines is likely to be cultural. Let us now turn to mathematics achievement in Australia for students in Years 11 and 12.

Recent Australian evidence of mathematics achievement in Years 11–12

Both IEA Mathematics Studies reported earlier indicate that, in each participating country, boys outperformed girls on the tests given to Year 12 students. They also make it clear, however, that the gap between male and female performance closed considerably between 1964 and 1978. There was no mechanism in the international study for statistically controlling for differential course taking. Nevertheless, the necessary information was collected in many countries, and regression analyses of the Australian data were used to investigate the effect on scores on the mathematics test of a variety of factors such as the sex of the student, the hours of mathematics instruction (some students will have taken one mathematics course and others two) and social class as indicated (predictably) by father's occupation (Rosier 1980; Moss 1982). The conclusions were that

> the main factors explaining differences in achievement between individual students were the amount of time spent in class on mathematics by the students, and hence their increased opportunity to learn the type of mathematics covered by the test items. The sex of the student was not directly associated with mathematics achievement, although male students tended to have higher ratings on the two main explanatory factors. (Rosier 1980, p. 167)

The IEA Mathematics Study tested every student enrolled in any mathematics course in Year 12 on the same test. Consequently, students were required to answer questions on, for example, calculus, regardless of whether or not they had studied calculus. Clearly, the subjects that are studied must influence the scores on such tests. It is arguable that course enrolments were different for boys and girls because of differences in their achievement levels, but the IEA data is incapable of providing evidence of the matter. To make inferences about the 'native' ability of any group on the basis of such data is inappropriate. We know that, in Australia, given the same formal preparation in mathematics, girls and boys achieve similarly until Year 10. Do differences develop as the nature of the mathematics that is studied changes during Year 11?

It appears not. Matriculation results in several states suggest that girls do very well in mathematics examinations at the end of Year 12. For example, in South Australia (*Education Gazette*, South Australia, 1979) girls' results in the 1978 external examinations were better (a greater percentage passing and also receiving A grades) than boys' on Mathematics 1S (a lower level matriculation subject) and on Mathematics 1 (a higher level matriculation subject) and equal to boys' on Mathematics 2 (a second higher level matriculation subject). This has also been the pattern during the 1980s. Similarly, Parker (1984, p. 4) comments of mean examination scores: 'Compared to boys it would seem that girls in Western Australia do very well in mathematics. The only significant differences which occurred were all in girls' favour'. In both South Australia and Western Australia, fewer than one-third of the students in the higher level mathematics courses were girls. A sensible interpretation of these and the IEA data is that girls enrol in mathematics courses which are easier than their ability would indicate to be appropriate. Consequently, a relatively more able group of girls is being assessed in each case. (We cannot, by the way, accept this explanation of boys' relatively poorer performance in the examinations in these states and still hold that the lower levels of participation of girls than boys in higher level mathematics courses reflects their relative abilities!)

The weight of Australian evidence, then, is that when 'opportunity to learn' is equal for boys and girls, so too are overall levels of achievement. This generalisation, however, masks an important feature of the achievement data. Small but consistent differences appear at both extremes of the achievement range, with boys being over-represented amongst the very highest and the very lowest achievers (Parker 1984; Parker & Offer 1987).

The evidence of gender differences in the extremes of mathematics achievement

In 1980, Leder reported that in the period from 1970 to 1979 relatively more Victorian boys than girls achieved the top grades in the highest level courses in Year 12 (Leder 1980b). Parker (1984) also found that, while on average girls' achievement in the highest level courses in Western Australia equalled or exceeded boys', fewer girls were represented in the top 1% of the achievement range of students enrolled in the courses. Similar patterns of achievement are observed in South Australia.

Talent searches also 'find' more boys than girls. For example, Kissane (1986) reports from Western Australia that, of Year 8 children nominated for academic extension programs, there are more very high-scoring boys than girls. In 1983, in the Australian Mathematics Competition (which one in five secondary school students and equal numbers of boys and girls enter), as early as Year 7 the ratio of boys to girls in the top 5% of the entrants (and, therefore, in approximately the top 1% of the Year 7 population) was 2.3:1 (Edwards 1985). At Year 12 level, the differences were much more extreme in favour of boys, although they were also more difficult to interpret because of the gender imbalance in entrants. Edwards points out that, in this competition, boys on average give only one more correct answer than girls; none the less, the effect on prizes is considerable. To win a prize at all one has to be in the top 0.5% of *entrants* and, in 1983, the ratio of winning boys to winning girls was 3.8:1 for Year 7 and 17.3:1 for Year 12.

At the other end of the spectrum, it seems that boys are more often found in remedial classes in mathematics. Again, how one interprets this is not clear. It may well be that, for a range of reasons, boys are represented to a greater extent than girls in the least able groups in mathematics. Equally, it may be that parents and teachers regard mathematics as more important for boys and, as a consequence, are more likely to recommend that they receive extra help when it is needed. At the present time, this is an area of considerable speculation.

If these differences at the extremes of the ability range are real (which is not to say innate) then some inconsistencies in achievement data may be explicable. For example, it is likely to be the students who are achieving least well who leave school earliest and, given the nature of the labour market, these are also more likely to be boys. Consequently, with the 'tail' of least well achieving boys removed, the average achievement of boys compared to that of girls will increase. For similar reasons, while even in the lower secondary years broadly based tests may indicate similar average achievement for boys and girls, talent searches, which focus only on the more mathematically successful students, suggest superior achievement for boys.

In conclusion, it appears that in many (although certainly not all) countries and cultures some differences in mathematics achievement do exist, but typically they are small and do not always favour boys, do not appear at all age and ability levels and are not consistent for all types of mathematics learning or all mathematics topics. In Australia, when participation is equal, achievement levels are likely to be as high for girls as they are for boys except in the top 1% of the mathematics achievement range. Regardless of the extent or cause of differences in 'mathematical giftedness', most girls achieve at least as well in mathematics as most boys. Consequently, the major focus of attention has moved away from girls' achievement in mathematics to their participation.

The evidence of gender differences in participation in mathematics

Little has been published in Australia on participation rates in mathematics in the compulsory years of schooling (Years 1–10). While in all states some study of mathematics is required during the early secondary years, the states differ in their degree of compulsion with regard to the level and hours of mathematics that is studied and in the degree to which the particular mathematics courses that are taken may be at the discretion of students (and their parents), the school or the system. Thus, as previously indicated, in Western Australia (until 1987) the level and hours of mathematics studied by a student in Years 8–10 was typically a school decision which was in turn based on state-wide comparability tests. This changed in Western Australia in 1988 with the introduction of the unit curriculum, which gives students considerably more choice over the amount and level of mathematics that is studied. In Queensland and New South Wales, all students study mathematics until the end of Year 10 and the level at which students study is almost always based on the school's recommendation. In other states and territories, while it is possible for students to choose their own mathematics units, it is unusual for students to discontinue the study of mathematics before Year 10.

Jones (1988) has provided detailed data on participation in mathematics and science in the final two years of secondary education in Australia. This report indicates that there are considerable variations among the mainland states of Australia in both male and female participation in mathematics. As Table 1 shows, retention of Year 8 students into Year 12 mathematics varies by mainland state from 55% to 16% for females and from 50% to 21% for males. Dekkers, De Laeter & Malone (1986) provide 1985 data for the Australian Capital Territory (ACT) and Tasmania which suggest that the corresponding figures from the ACT are higher and from Tasmania lower than those of any of the mainland states. In each case, this reflects the overall retention rate from Year 8 to Year 12 (highest for ACT and lowest for Tasmania). In New South Wales, Queensland and Western Australia, over 90% of students who reach Year 12 take at least one full year of Year 12 mathematics. By comparison, in Victoria and South Australia, the overall proportion of Year 12 students taking at least one mathematics course is about one-half. More particularly, where male participation is high the ratio of female to male enrolments also tends to be high. It may be that in those states the study of mathematics is most taken for granted and the timetabling of upper school courses assumes the study of mathematics. (In the ACT in 1985, participation in a mathematics course as a percentage of Year 12 was about 75% for both boys and girls, and the ratio of girls to boys was approximately 1).

Table 1
Students studying Year 12 mathematics courses in 1986

	Percentage of Year 8 cohort Females	Males	Percentage of Year 12 Females	Males	Ratio of females to males
Qld	55	50	90	92	1.07
NSW	44	42	94	99	1.00
WA	32 (44)	39 (47)	61 (84)	81 (97)	0.86 (0.96)
SA	21 (29)	28 (35)	36 (50)	54 (69)	0.74 (0.82)
Vic.	16 (21)	21 (25)	31 (40)	50 (59)	0.83 (0.89)

Note that for Victoria, South Australia and Western Australia two sets of data are provided. The figures not in brackets represent the percentage of students enrolled in externally examined or moderated Year 12 courses; the figures in brackets include the school-assessed courses which are not used for the calculation of tertiary entrance scores (TES). While within any particular state the latter are likely to be somewhat easier than the former, the same assumption cannot be made across states, since whether or not a course is used for TES depends on state tertiary admission structures. For example, in Western Australia, Mathematics IV could be used for TES in 1985 but not in 1986. However, it was the tertiary admission structure which changed in 1986 and not the course.

The data for the five mainland states have been compiled from Jones (1988) which provided no information for Tasmania and the territories. The percentages are based on course enrolment ratios of males and females obtained through telephone calls to relevent bodies. They are consistent with the information provided for the National Data Base on Girls and Education which is yet to be published.

It would be interesting to speculate on the relationship between the level of compulsion associated with the study of mathematics in the lower secondary school and the participation rates in the upper secondary school. Indeed, one is tempted to suggest, contrary to popular thinking on the matter, that compulsion in the lower secondary school does not simply delay the decision of girls to withdraw from the study of mathematics but actually influences it. It may be that early adolescence is a particularly inappropriate time for girls to have to make subject choices. This, however, is still speculative and, as will become clear from Table 2, girls generally do participate in the highest levels of mathematics to a lesser extent than boys.

Table 2
Students studying a highest level Year 12 mathematics course in 1986

	Percentage of Year 8 cohort Females	Males	Percentage of Year 12 Females	Males	Ratio of females to males
Qld	6	15	10	28	0.37
NSW	7	13	16	30	0.56
WA	4	12	8	24	0.40
SA	7	14	11	27	0.48
Vic.	4	10	8	24	0.42

Note that the data for the five mainland states have been compiled from Jones (1988) which provided no information for Tasmania and the territories. The percentages are based on course enrolment ratios of males and females obtained through telephone calls to relevent bodies. They are consistent with the information provided for the National Data Base on Girls and Education which is yet to be published.

About twice as many boys as girls take advanced mathematics nationally, although the state and territory ratios of males to females vary from 2:1 to 3:1. Girls, it seems, are less likely to study higher levels of mathematics. Often students studying a highest level mathematics course are required to study two courses or a double course. The direction of the relationship between these factors is not clear; that is, it is not clear whether girls are reluctant to study double mathematics and, consequently, do not have access to the highest level mathematics or whether the reverse is more common. Either way, the result is that the mathematical 'yield' (Keeves & Mason 1980; Moss 1982; Dekkers, De Laeter & Malone 1986) is lower for girls than for boys, and herein lies the source of the current focus of interest in the problem of girls and mathematics—why *won't* girls do as well as boys in mathematics? From the plaintive way in which this question is often asked it would seem that many believe the girls to be simply perverse, not prepared to be told what's good for them and take the (admittedly unpleasant, some would say) medicine they need; others regard the girls as compliant victims participating in their own oppression.

Although in most Australian states girls are increasingly taking mathematics in Years 11 and 12, attempts to convince girls to participate in the highest levels of mathematics have not met with overwhelming success. Many still believe that females are 'naturally' less able and/or interested than males in mathematics and that this is the reason why they choose to participate less. Others reject this claim, considering it ahistorical and arguing that, just as achievement differences have virtually disappeared, so too can differences in levels of participation. Unfortunately, community beliefs about gender and achievement and participation in mathematics are difficult to change and are likely to have a powerful influence on girls' and boys' achievement and participation. For example, what is likely to be the impact on girls and their families, teachers and peers of the oft-quoted evidence that boys are more likely to be mathematically gifted than girls?

Questionable uses of the available evidence

As suggested earlier, it is regularly found that more boys than girls are represented in the very highest achievement bands in tests used to select mathematically gifted children and amongst prize winners in mathematics competitions (e.g. Australian Mathematics Competition, International Mathematics Olympiad, exhibitions for outstanding achievement). In the United States of America in 1980, Benbow & Stanley argued that, since as early as 12 years of age more boys than girls exhibit extreme precocity on the mathematics subtest of the Scholastic Aptitude Test, and since at this age formal mathematics experiences will be similar for boys and girls, we can infer that girls have poorer mathematical aptitude. Benbow & Stanley (1980) expressed a preference for a genetic explanation of this observed difference and further argued that this evidence supports the conclusion that 'less well-developed mathematical reasoning contributes to girls taking fewer mathematics courses and achieving less well than boys' (Benbow & Stanley 1980, p. 1262). Thus, they argued that the evidence from the talent quests implied that lower participation by girls in mathematics was caused, essentially if not completely, by girls' lower aptitude.

Their article was published on 12 December 1980 and attracted immediate media attention in the United States of America and also in Australia:

'Do males have a math gene?' (*Newsweek*, 15 December 1980)
'The gender factor in maths' (*Time*, 15 December 1980)
'Male superiority' (*The Chronicle of Higher Education*, December 1980)
'Boys have superior math ability, study says' (*Education*, USA, 15 December 1980)

It proved extremely controversial for a number of reasons. As Fennema (1981) pointed out, what is the value of research when this one study, which provided no justification for a 'preference' for a genetic explanation for the differences that were observed and which almost completely ignored previous research on the matter, was given such a high profile both in the popular and in the educational press? Seven of the eight replies to Benbow & Stanley, published in the April 1981 edition of *Science*, were critical of the conclusions on a wide range of grounds, as were several later publications. These rebuttals, however, received no such publicity. More recently, during the full week of the Australian and New Zealand Association for the Advancement of Science (ANZAAS) conference (May 1988), the only conference paper to 'hit the headlines' of the national newspaper, the *Australian* of 18 May 1988, was the one which presented updated information on the results of the Australian Mathematics Competition. Given the evidence presented in this monograph so far, both the title of the front page article, 'Girls don't count in maths', and the assertions therein were staggering:

> More boys than girls can recite Don Bradman's batting average and Dennis Lillee's bowling record — but it may not just be because of their obsessive interest in sport.
> When it comes to figures, girls just don't count. In fact, boys are more proficient than girls over much of the spectrum of mathematical disciplines.

Furthermore, the ANZAAS conference paper authors were misrepresented and were dismayed at the interpretation placed on their data and at the snide use to which it was put.

The constant male dominance of mathematics talent quests and prize givings and the media attention given to any suggestion of intellectual (and sporting!) differences between males and females are likely to have a powerful impact on community perceptions of females' mathematical talent. Since we do not award prizes for low levels of achievement, boys' over-representation amongst the mathematically least able does not receive much media or educational coverage — the popular message is that most boys are more able mathematically than most girls. Thus, as Yates & Firkin (1986, p. 18) suggest, 'the mythology tends to ascribe the outstanding competitive achievement of the minority of gifted boys to all boys as a general attribute of masculinity'. As has already been shown, however, and contrary to the view expressed in the *Australian*, the facts are that many girls achieve better in mathematics than most boys and that many boys achieve less well in mathematics than most girls. Furthermore, and contrary to Benbow & Stanley's (1980) claim, differences exhibited in giftedness cannot directly explain the *extent* of differential participation in higher level school mathematics or in the large range of occupations which involve the use of mathematics. It may be, however, that these differences in educational and occupational participation can be explained in part by perceptions of males as typically more mathematically able than females. If perceptions of much of the community are still that girls cannot, need not and do not want to do advanced mathematics (or at least not to the same extent as boys), then these views may also indirectly influence girls' achievement in mathematics. For these reasons, it is important that we understand why males are typically more heavily represented in the highest achievement bands, and in the following chapter I detail some of the explanations that are offered. Before proceeding, however, I would like to make two points.

Firstly, there is a danger that in focusing on this very narrow range of students and this particular form of competitive success one gives added credence to the thoughtless and unwarranted, if not grossly irresponsible, views such as those expressed in the *Australian*. Reports such as these can undo years of work undertaken to improve girls' real access to mathematics and one is tempted to dismiss them out of hand, but there are clear dangers in doing so. Let me emphasise that the proportion of this monograph dedicated to a discussion of the evidence of giftedness reflects the political reality that this evidence is used, one would like to believe inadvertently, to undermine girls' success in the so-called 'masculine' endeavours, and I include it because the issue must be addressed.

Secondly, while Australian research directed at understanding gender differences in participation and achievement in mathematics is growing rapidly, it is still limited compared to the amount of similar research in the United Kingdom and the United States of America. Consequently, in what follows, I have relied somewhat on research undertaken elsewhere. Notwithstanding the reservations expressed earlier regarding the unquestioning application of international data to Australia, such research can be valuable in understanding local findings, in suggesting directions for study and, when consistent with local findings, in strengthening our confidence in them.

Chapter 3
Explanations of gender differences in giftedness in mathematics

Benbow & Stanley (1980) expressed a preference for a genetic explanation of differences in mathematical precocity although their study provides no evidence which can help decide the issue and offers no explanatory mechanism apart from a suggestion that spatial visualisation may be implicated. During the 1960s and 1970s a rather extensive literature developed around possible genetic mechanisms for differences in spatial and mathematical abilities and, certainly, theories abound (for reviews and critiques of some of the models that were offered see Harris 1978 and McGee 1979 about spatial ability and Widdup 1980 and Atweh 1980 about mathematical ability). Those concerned with educational practice, however, are not likely to find such debates fruitful since they provide little advice about appropriate action. As a group of six scientists replying to Benbow & Stanley stated:

> An underlying fallacy . . . largely responsible for the unwarranted publicity in the popular media . . . is the notion that, if a trait is under genetic control, the expression of that trait is immutable . . . One has only to consider something as simple as the height of wheat to realize that a particular genetic strain of wheat will yield different ranges of height in different environments. A determination of the height in one environment will tell one nothing about the height in a substantially different environment. (Egelman, Alper, Leibowitz, Beckwith, Levine & Leeds 1981, p. 116)

This raises the important question of the conditions that are most supportive of high levels of mathematics achievement for girls.

Some cross-cultural evidence
It does appear that, in most countries, more boys than girls are amongst the very high achievers in mathematics but, even within this very narrow achievement band, the extent of the differences greatly depends upon the children's home culture. For example, in a talent search for mathematically gifted children in Rhode Island, USA, 17.5% of the winners were Asians even though only 1.7% of the population of the region were Asian. More particularly, of the Caucasian winners only 27% were girls, while of the Asian winners 47% were girls (Campbell & Connolly 1984).

When coupled with the clear differences in childrearing patterns that were observed amongst these Asian and Caucasian families, this provides quite powerful evidence that there can be no simple genetic explanation for gender differences in giftedness. For example, Campbell & Connolly found that the Asian students generally spent twice as many hours as the Caucasian students on study, their work was more closely directed and supervised at home, and they were also very group oriented and strongly influenced to enrol in advanced mathematics courses by peers and parents who had high expectations of them. Asian girls had more technical books and knew more computer languages than all the other groups. In the Asian families, the prestige and honour of being selected in a mathematics talent quest were regarded as very important for both boys and girls. Caucasian students, by contrast, were expected to take more charge of their own work and to be independent. The parents of Caucasian boys stressed the importance of productive technical careers for their sons, while the parents of Caucasian girls emphasised 'being happy and well adjusted' as a primary goal for their daughters. Also, the Caucasian parents were less happy for their daughters than for their sons to participate in the research classes that were involved in the enrichment program because the classes required attendance at laboratories out of school time, necessitating that the girls travel alone. These differences were not found amongst the Asian families. Caucasian boys, Asian boys and Asian girls all rated money higher than satisfaction in career choice, whereas the reverse was true for Caucasian girls. Differences in the stereotyping of mathematics, science and technology as male domains were also quite marked among the Asian and Caucasian parents, girls and boys. The Asian boys had few negative perceptions of the gifted girls, whereas the Caucasian boys had many. In addition, the parents of the Asian girls considered that family networks would enable their daughters to balance careers and families. Other North American studies have confirmed similar attitude and achievement patterns; for example, in Hawaii, girls of *each* racial subgroup tend to outperform boys even in the upper achievement range (Brandon, Newton & Hammond 1985).

Equivalent Australian evidence is not available. Lovejoy & Barboza (1984), however, carried out a comparison study in New South Wales of Australian students and students from certain Asian countries (Hong Kong, Indonesia, Malaysia, Singapore and Taiwan) who had English and Chinese, respectively, as their first language. They found higher levels, and higher female to male ratios, of participation in mathematics-related university programs (commerce, applied science, science, engineering and management) amongst those with Chinese as their first language than amongst those with English as their first language. Further, gender differences in mathematics anxiety and attitudes towards mathematics were greater amongst the Australian students than amongst the Asian students. The authors suggest that the interpretation of mathematics and science as being male domains is specifically a Western phenomenon. The Chinese women did not appear to regard femininity as in conflict with mathematical thinking. The conflicts between intellectuality and femininity often associated with learning mathematics are not, it seems, related to some essentially conflicting quality of mathematics and famaleness, respectively. *If* gender differences in mathematical talent are genetically determined, then we can be reassured that biology is *not* destiny.

Indeed, the weight of the explanations lie with social–cultural explanations such as that girls and boys may differ in their 'opportunities to learn' mathematics, that girls may be less confident with the subject and suffer greater anxiety about the social consequences of exceptionally high attainment, and that the particular form a test takes is likely to be a telling feature in determining whether or not gender differences in achievement appear. I will consider each of these in turn.

Opportunities to learn mathematics

It is often suggested that the types of toys provided for girls and boys differ considerably (Rheingold & Cook 1976), with girls being given dolls and boys scientific and spatial toys, and that the play and other kinds of activities in which girls engage do not provide them with experiences appropriate to the development of spatial concepts and skills thus inhibiting their achievement in higher levels of mathematics. Many who do not accept the biological determinism which holds that girls are 'naturally' (that is, genetically) less mathematically able than boys do hold to a kind of cultural determinism which suggests that separate female and male cultures are 'natural' and that girls and boys, as members of these respective cultures, differ in their interest in mathematics and a range of associated activities. From either perspective it is generally held that girls have the same choices as boys but for one reason or the other (or both) do not choose to study mathematics formally and informally to the same extent as boys and, consequently, underachieve in mathematics. Young children's play and older children's hobbies are offered in support of this view ('that's just what girls and boys are like'). My purpose here is not to argue the case for or against cultural determinism, but I will point out that all cultures are dynamic and that both at home and in school we work extremely hard, if not always consciously, at constructing gender (Kenway & Willis 1986).

Whether one accepts a deterministic position or not, there is evidence that girls' and boys' play differs along several dimensions (Tittle 1986). That 'girl's play' is any less spatially complex than 'boy's play', however, has been disputed (Gelman & Carey 1981; Walden & Walkerdine 1982). It seems more likely that preconceived and stereotyped notions about the appropriateness of certain forms of play (in particular, actions on objects) for the development of mathematical processes has led to the dismissal of girls' play as inappropriate or unproductive for their mathematical development.

In any case, while there is evidence that girls' spatial ability tends to be less well developed than boys' (Wattanawaha & Clements 1982) and while the connections between geometry and spatial skills seem obvious, the relationship between spatial development (and the term 'spatial' is used to cover a wide range of skills) and achievement in the broad spectrum of mathematics is unclear (Fennema & Tartre 1985). Fennema has researched the role of spatial visualisation in mathematics achievement for more than a decade. In the conclusion to a recent study, she and a colleague stated that 'one should question the idea that spatial visualization skills are highly important in the learning of mathematics and that the development of such skills should become a major goal of mathematics education' (Fennema & Tartre 1985, p. 203). None the less, they and others (for example, Sherman 1979; Ethington & Wolfle 1984) have suggested that spatial visualisation appears to have little effect on mathematics achievement for boys but a somewhat greater effect for girls. Gender differences in spatial visualisation skills do not appear, however, in groups of mathematically gifted children, and spatial visualisation does not correlate with mathematics achievement for these children (Weiner & Robinson 1986).

Differences in 'tinkering' experiences and in the kinds of hobbies and extra curricular activities in which girls and boys engage are well documented (Stage, Kreinberg, Eucler & Becher 1985; Handel 1986), however, and these may well influence girls' knowledge of many of the contexts in which mathematical ideas and problems are presented and hence their achievement in mathematics. Thus, consistent differences in geometry and measurement tasks may be a direct consequence of such experiential factors, not necessarily mediated by the development of spatial visualisation skills.

Increasingly, it is suggested that patterns of student–teacher interaction in classrooms occur along gender lines. For example, in the mid-1970s Leinhardt, Seewald & Engel (1979) videotaped 49 Year 2 teachers in the United States and found that they made more academic contact about reading with girls than with boys and about mathematics with boys than with girls. Although the differences were modest, Eccles & Jacobs (1986) estimated that if the pattern were maintained the difference in teaching time on mathematics could amount to as much as 36 hours by the time the children reached Year 7. This is equivalent to a school year of mathematics—a substantial difference in girls' and boys' opportunity to learn mathematics (and also reading, of course!). More recently, and in Australia, Leder (1986) videotaped teachers of Year 3 and Year 6 classes and found that the teachers consistently interacted more frequently with boys than with girls in *both* reading and mathematics, with boys initiating many more interactions than girls. In particular, boys received both more disciplinary interactions (encouraging them to be task oriented) and more process questions (for example, requiring synthesis or generalisation).

While it is clear that a particular student's opportunity to learn mathematics directly from her or his teacher is less if that student has fewer mathematical interactions with the teacher, the direct influence of teacher time on student achievement is unclear. It may well be that the amount of interaction between student and teacher has less to do with achievement than has the nature and pattern of that interaction. In the United States of America, Fennema & Peterson (1986) and Peterson & Fennema (1985) studied student–teacher interactions and the relationship between student–teacher interactions and low-level and high-level mathematics achievement. They found that there were some subtle gender-related differences in types of interactions but, more often than not, teachers responded similarly to boys and girls; that is, they praised, prompted, criticised and corrected, and gave and refused help in similar ways for boys and girls. None the less, there were more interactions with boys than with girls. This was so for all categories of teacher- and student-initiated interactions, and for mathematical, procedural, behavioural and social interactions, and was significant for several. In particular, teachers initiated significantly more behavioural, procedural and social interactions and more high-level and low-level mathematical interactions with boys than with girls. Teachers more often ignored girls' requests for assistance and less often accepted their 'called out' answers. They also interacted more with high-achieving boys than with high-achieving girls, and with highly confident boys than with highly confident girls.

A particularly interesting finding was that different patterns of interaction seemed to relate to high and low levels of achievement for boys and girls, respectively, and for achievement of high- and low-level learning. For example, teacher-initiated contacts which involved calling on a student who did not have his or her hand raised (non-volunteer) to respond to a low-level mathematical question were positively related to success for boys on high-level tasks, but negatively related to success for girls on high- and low-level tasks and for boys on low-level tasks. Conversely, calling on non-volunteer students with high-level questions was related to high-level mathematics success for girls but did not affect achievement for boys. A range of other similarly complex patterns was observed, indicating that levels of teacher interaction do not have an easily predicted influence on achievement. In particular, the impact of student–teacher interactions on high-level learning is not always the same for girls and boys and, for both boys and girls, the interactions which may promote low-level learning may inhibit high-level learning.

Clearly, there are no simple rules for effective gender-inclusive teaching, and simplistic recipes relating teacher-time to student achievement are likely to be unproductive. The preceding studies were unable to illuminate the causes of the patterns of interactions that were observed and, obviously, teachers' intentions are central to an understanding of what happens in classrooms. In any case, students are not passive recipients of this process; they also have a role in moulding the form and content of classroom interactions. Fennema & Peterson (1986) do suggest, however, that their data provide support for an earlier hypothesis (Fennema & Peterson 1985) that teaching strategies which encourage students to engage in autonomous learning behaviours assist them with learning high-level mathematics.

Confidence, anxiety and success in competitions

The term 'autonomous learning behaviours' refers simply to the ability to work independently, and to persist with, choose and succeed at high-level cognitive tasks. Fennema & Peterson (1985) suggest that the only way one learns to do high-level tasks is by choosing, persisting and succeeding at them (!) and that, for a variety of reasons, girls do not develop these behaviours in mathematics to the same extent as boys. The argument is that girls more often than boys attribute their failures in mathematics to lack of ability and their successes to external features, such as luck or an easy test, or to hard work and, consequently, doubt their capacity to avoid failure or achieve success. Opposite patterns of attribution are often reported for boys who, it is suggested, are more likely to attribute failure to lack of work. (It should be noted that while the empirical evidence on the whole supports this claim, it is not completely consistent: compare Clarkson & Leder 1984; Leder 1984; Leder 1985.) As will be discussed later, girls compared to boys also exhibit less confidence in continued achievement in mathematics and are less convinced that mathematics will be useful in their future lives. Some girls may also see some conflict between being able to 'think mathematically' and being female. These factors, together with the gendered patterns of interaction in many classrooms, may inhibit the development of autonomous learning behaviours (Fennema & Peterson 1985).

Dweck (1986) has suggested that such differences in self-perceptions and associated behaviours may contribute to the discrepancies that are found between boys and girls in high levels of mathematical achievement. On the basis of considerable research by herself and others, she concludes that

> ... bright girls compared to bright boys (and compared to less bright girls) seem to display shakier expectancies, lower preference for novel or challenging tasks, more frequent failure attributions to lack of ability, and more frequent debilitation in the face of failure or confusion ... (Dweck 1986, p. 1044)

and that these behaviour patterns are likely to produce greater declines in confidence for bright girls than for any other group when confronted with unfamiliar ideas requiring new conceptual frameworks. According to Dweck, the causes of such patterns of behaviour are complex, but it appears that children who are oriented towards 'performance' criteria, the goals of which are to gain positive judgments and avoid negative judgments of what they consider to be a fixed capability or competence, are more likely to suffer from such counterproductive behaviour patterns when confronted with the unfamiliar than children who are oriented towards 'learning' criteria, the goals of which are to increase competence. Dweck (1986, p. 1046) suggests that children who have had 'early, consistent and abundant' success but who have been unchallenged and protected from failure and have come to attribute their failures to lack of ability and their successes to features such as behaviour, neatness or carefulness often fear the prospect of challenge in mathematics because they are unsure of continued success.

Girls, in particular, have 'early, consistent and abundant success' with computation, and, although competition may be prevalent in the teaching of computation, challenge rarely is. Furthermore, as I have argued elsewhere, error making and risk taking are typically not regarded as a constructive part of the mathematical learning process (Willis 1985). Failure is often seen as somehow nasty, to be blamed either on the child for lack of effort or on the teacher for not planning sufficiently carefully. The thing to do is to reduce the risk of error and, in the short term at least, the way to reduce error is to reduce risk taking. Indeed, in many classrooms a caring and good teacher is regarded as one who carefully sequences and structures the mathematics curriculum so that children can almost always respond correctly. And there is evidence that girls more than boys are protected from such risks (Fennema & Peterson 1986; Leder 1987). According to Dweck, however, research

> ... is clear in indicating that continued success on personally easy tasks (or even on difficult tasks within a performance framework) is ineffective in producing stable confidence, challenge seeking, and persistence ... Indeed, such procedures have sometimes been found to backfire by producing lower confidence in ability ... Rather, [better] procedures ... are the ones that incorporate challenge, and even failure, within a learning- [as compared to performance-]oriented context ... (Dweck 1986, p. 1046)

Looked at from a slightly different perspective, the anxiety associated with a new, possibly insoluble problem is a part of doing mathematics (and any other creative activity). The task is not to remove anxiety but, rather, to recognise its role in producing that essential adrenalin and to manage it so that it does not become incapacitating. Many people have an image of the 'mathematical mind' and believe that those who have such a mind can complete mathematical tasks quickly, easily and at the first attempt. (This view is, of course, promoted by timed examinations and competitions which together trivialise mathematical thinking.) As those who have done mathematics to any extent will testify, this perception is absurd, but it is likely to have a powerful influence on children and adults when they confront unfamiliar mathematical tasks. Some children, more than others, are protected from mathematical risks and consequently are prevented from learning to 'flounder' in a constructive way (Hunt 1985). If these children have read all the implicit messages in being so protected (and many other associated messages, of course) and have come to interpret an inability to see an immediate solution to a mathematical problem as evidence that they cannot solve the problem or, worse, do not have that elusive 'mathematical mind', then perhaps they are more likely to be stressed when confronted with problems to which they can see no immediate or obvious solution and, consequently, may give up on them. As has been suggested, there is a great deal of evidence that such children are more frequently girls than boys, but there is also evidence that they are more frequently working-class children than middle-class children and more often black than white (Reyes & Stanic 1988). Mathematics, it seems, is the province of middle-class males, particularly those of Anglo-Celtic and sometimes Indo-Chinese extraction.

It has been suggested that those who become successful in mathematics and mathematics-based careers are those who resisted school mathematics—who have, in essence, taught themselves mathematics (Easley & Easley 1982). The 'successful' student must adopt the framework of the discipline—the ways of seeing and saying—yet, even at the upper levels of mathematics in the secondary school, these are quite in opposition to the narrow socialisation of much school mathematics. The implication is that many who have been well socialised to mathematics in the primary school may suffer the consequences as they attempt to move on to more sophisticated mathematics and that these children are more often able girls.

> ... many girls achieve very strong schema for the algorithm-dominated mathematics of the top primary school. When the demands of mathematical learning change in the secondary schools the early successful schema are inflexible and make the necessary accommodation difficult. Boys tend to be less effective at upper junior school algorithmic mathematics and thus develop less inflexible schema. (Woodrow 1984, p. 7)

The argument is that it is *because* girls become so good at computational mathematics that they experience some difficulties with higher level mathematics. Paradoxically, it seems that girls' very success at learning primary school mathematics is turned against them. Is it that girls are 'naturally' better at rote algorithmic tasks, or is it that the very process of learning computation so well 'ruins them' for other 'high-level' mathematics. The work of Bourke & Stacey (1988), amongst others, would suggest not. How do we explain this apparent contradiction?

Walkerdine has studied British primary and lower secondary mathematics classrooms extensively. She argues that child-centred learning theories 'present children's natures as gender-neutral and universal, but, when it comes down to it, deeper examination reveals just how class, race and gender specific this supposed nature is' (Adams & Walkerdine 1986, p. 14). The language used to describe children, in general, is that most closely associated with masculinity rather than femininity, the latter being more closely associated with the more traditional ways of teaching—passivity, rote learning and rule following (Walkerdine 1983). Within child-centered learning theories, the 'natural child', that is, the 'active', 'inquiring' child, is typified as the child with all the characteristics associated with a white, middle-class male. Boys' play, it appears, is more 'active', more 'exploratory', more 'outward pushing' and more 'constructive', and involves more 'prediction and planning', while girls' play is less 'constructive', more 'structured', more 'rule oriented' and more 'inward looking', has more guidelines for 'behaviour', and involves more 'fantasy' (see, for example, Shuard 1982; Taylor 1986; Tittle 1986). Even today, boys who behave in active, enquiring ways are 'real boys', but girls who behave and may succeed in these ways are 'tomboys', possibly even 'aggressive' (Walkerdine 1983). Girls who succeed in other ways (who do not fit our image of the active, enquiring child) somehow are not 'really capable'; *girls' success*, as was remarked earlier, *must be explained away*.

As we have seen, girls do very well at mathematics in the early primary years but such early success turns out to be no success at all. Often on the implicit grounds that it cannot be 'natural', girls' early learning is not regarded as *real* learning because they have learned *'in the wrong way* . . . Instead of thinking properly, girls *simply* work hard' (Walkerdine 1983, p. 84). Conversely, at least for middle-class boys, early failure turns out to be no failure at all. Such boys make mistakes because they are bored, or they may not be as mature as the girls, but they have potential. Girls who are succeeding work hard, plod, are average and nice, capable and try hard, are kind, helpful and work quietly, and are either confident or lack confidence; boys show true understanding, are mathematically minded, bright, bored, have potential, are creative, difficult, obstreperous, naughty and dynamic (Adams & Walkerdine 1986). Which set of descriptions more closely fits our image of a future mathematician?

The social typification of girls' success in mathematics as due to hard work and rule following can encourage the kind of caring and protective behaviour on the part of teachers which leads girls to believe that they cannot *really* cope. Thus girls learn to believe that their success must be of a particular, less worthy kind. Walden & Walkerdine (1986) suggest that contradictions between the practice of school mathematics, which is about rule following, and perceptions of the discipline of mathematics, and hence the nature of *real* mathematical learning, as being about rule challenging, highlight discontinuities in the ways that girls' femininity is defined, described and develops. Girls' poorer performance and increasing anxiety as they proceed through school, Walden & Walkerdine suggest, is produced by these practices.

> Girls can be successful in terms of mathematical attainment, gaining power by taking responsibility in the classroom, but remain relatively powerless in terms of teachers' judgements of their performance. Since the latter depend on indications of the challenging of rules which are understood as 'real understanding', 'flair' or 'brilliance', girls are often left in an ambiguous position. (Walden & Walkerdine 1986, p. 143)

An important point to make here is that these arguments do not mean that the process is linear or unidirectional in the way we typically regard a 'self-fulfilling prophecy' to be, or that it is a simple process of socialising passively accepting girls (and boys) into expected roles. As was remarked earlier, students are *not* passive in this process; they struggle to make sense of conflicting messages. Many succeed in making a kind of sense of it, others resist in a range of ways. Furthermore, as Walden & Walkerdine (1986) warn, the argument should not be read as suggesting that girls are forced out of mathematics by a kind of patriarchical conspiracy. The contradictions of rule keeping and rule breaking are part of a whole constellation of classroom practices in which girls and boys and teachers are active and which, together with particular conceptions of the nature of mathematical learning (as compared, say, to English or social studies learning), produce gendered patterns of achievement.

Indeed, there is evidence that girls' success in mathematics in secondary school is also explained away and typified as due to hard work, attention, co-operativeness and neat presentation of work (Parker & Offer 1987), that, given girls and boys of equal mathematical achievement, the girls are less likely than the boys to be nominated for academic extension programs (Kissane 1986) presumably on the basis of something called 'potential', or to be placed in the highest ability group and also the lowest ability group in classrooms (Hallinan & Sorenson 1987). Thus, it is clear that mathematics is regarded as more important for males; mathematically 'weak' students who are male receive more remedial help and mathematically 'strong' students who are male receive more extension help than do similar females. Amongst equally achieving boys and girls, girls are often given explicitly different advice about subject choice in mathematics and about post-school options (Stanworth 1981), even when the person giving the advice assesses the students' current achievement levels accurately. It does appear that, in a range of ways, girls are silenced in classrooms while boys' experiences, concerns and viewpoints are validated (see further, Evans 1982). The messages many girls receive about their potential and their needs may be implicit, but they must also be 'loud and clear'.

Earlier it was suggested that differences in mathematics achievement in favour of boys are often exaggerated. Where they appear at all, it is likely to be at the highest level of mathematics and the differences are usually small. None the less, it is arguable that such small differences in achievement as do appear (and also the rather more significant differences in participation) can be explained in terms of the gender-differentiated production, in many homes and mathematics classrooms, of anxiety and doubt regarding 'intellectuality' of the kind associated with mathematics. Indeed, there is quite extensive anecdotal evidence that some intelligent and successful women experience strong feelings of guilt about the regard in which they are held, because they *know* that they are not really able to do mathematics.

> You see I always think people think too well of me and it is important to me that they shouldn't like me for things that I believe I do not possess . . . I have to go out of my way to prove that while I'm quite intelligent, I'm not *very* intelligent . . . *it is important that they know that I can't add up.* (quoted in Buxton 1981, p. 134)

We will return shortly to the moral overtones often associated with success and failure in mathematics. For the moment let me simply say that there is a real conflict here about what is acceptable with regard to mathematics. On the one hand, it is almost socially acceptable not to be able to do mathematics but, on the other hand, it is somehow impossible to be really intelligent if one is not good at mathematics.

The nature of the test and the testing situation

There is a compelling body of evidence to show that, irrespective of previous attainment, a person's confidence of success in a stressful testing situation will influence that success. For example, Adams (1986) has demonstrated that results in the Australian Scholastic Aptitude Test (ASAT) are strongly influenced by confidence in success in the test and that boys are significantly more confident than girls of such success even though previous results do not predict a gender difference in achievement. Confidence about success in ASAT does not influence success in other forms of assessment such as traditional examinations or school and university coursework, but it does influence success in ASAT itself. What is special about ASAT?

Firstly, ASAT is widely (but mistakenly) regarded as a measure of aptitude for tertiary study rather than a measure of previous achievement. Girls who attribute their success to conscientiousness and who have come to believe, with the woman quoted earlier, that 'while I'm quite intelligent, I'm not *very* intelligent' are likely to feel extremely threatened by situations which they believe rely on 'innate talent' or 'intelligence' rather than on achievement. As we suggested earlier, many implicitly (and even explicitly) believe in an intangible 'something'—called talent, potential or the mathematical mind—which some have and others do not and which is independent of previous learning or effort. If you think you do not 'have it', the feeling of waiting to be found out must intensify considerably in the context of competitions, tests of giftedness and other examinations which are regarded as assessing talent rather than 'simply' achievement. Given nothing more than the widespread belief that boys more often than girls have it, we are likely to find differences in boys' and girls' confidence of success in tests regarded by the students themselves as assessing intelligence or potential rather than achievement.

Secondly, it is suggested that many tests of achievement are not necessarily equally appropriate for girls and for boys. For example, Bourke & Stacey (1988) have demonstrated that the method of administration of a test can affect overall levels of achievement and gender differences. Achievement in a six-problem mathematics test was higher when the items were presented one at a time than when they were presented as a set. Furthermore, the improvement was greater for girls than for boys. It also appears that girls are generally favoured by essay type questions, boys by multiple choice questions and neither by structured papers (Eccles & Jacobs 1986; Slack & Porter 1980). If this perception is correct, then it has very signifcant implications for the validity of our assessments, given the preponderance of multiple choice test formats in mathematics competitions (e.g. Australian Mathematics Competition), in tests used to determine giftedness (e.g. ASAT and the US Scholastic Aptitude Test), and in national (ACER tests) and international (IEA) assessments.

Sassen (1980) offers another intriguing possibility. She is particulary critical of the 'success anxiety' and 'motive to avoid success' constructs which are often used to explain girls' and women's under-representation in the highest achievement bands in mathematics. It is not success that causes females anxiety, she suggests, but the very climate of competition. There is, indeed, growing evidence that the average level of achievement of girls is likely to be higher in co-operative learning structures than in competitive ones (Peterson & Fennema 1985). Sassen (1980, p. 18), in a discussion of success in the corporate world, suggests that women, more often than men, 'are unable to take competitive success and construct around it a vision, a new way of making sense, to which they can feel personally committed'. It is not girls and women who need to change, she claims, but rather, for the good of us all—boys and men included—the 'corporate environment' needs to be restructured so that competition is not the only avenue to success. Surely, we could expect no less of the 'learning environment'.

But what does any of this have to do with mathematics examination and competition results? The point is that success is unlikely to be independent of the way one feels about one's chances of success or of the impact of success on one's life. Any sports coach will testify that talent is not enough to produce 'a champion': a champion has to like to compete, want to win very badly and *know* that he or she is a winner, and that the prize is within grasp. We readily accept the powerful effect of 'hype' on high levels of achievement in sport. Is it likely to be less so in the intellectual sports? And, no matter how we dress it up, mathematics competitions and talent quests are competitions like any other and the prizes go, not necessarily to the 'best man'(!) but to the best competitor. Unfortunately for girls who might be exceptionally successful in mathematics, the associated hype is more likely to convince them that they *cannot* win.

In this chapter, we have been considering students who are considered gifted in mathematics. But, surely, real success in mathematics is not about competition. Mathematicians are rarely in races—are timed and competitive tests really a good measure of mathematical achievement, 'talent' or 'potential'? The idea seems ludicrous in the extreme. Mathematical thinking is as much about problem posing as about problem solving, whether one's first idea happens to lead to a solution is as much about luck as about talent and, certainly, mathematicians expect a great many 'dead ends'. Often the attempts that happen not to provide a solution to a particular problem lead to new problems; some lead to new theories and great advances. Surely, at every level, good mathematics is more about reflection than about speed. What do these competitions really measure? Would we find different winners if the competitions were more like those in the other arts and sciences, where the competitors set their own problem, provide the solution, determine when the task is complete and present the result in their own way?

Chapter 4
Explanations of gender differences in participation in mathematics

Although considerable research has been devoted to identifying the reasons why girls and women choose not to continue to study mathematics, the influences on decisions about subject and occupational choice appear to be sufficiently complex and subtle to suggest that many decisions are made without a conscious awareness of the contributing factors. This makes it rather difficult to obtain direct evidence of the factors influencing choices. In broad terms, however, the explanations seem to centre on girls' confidence and anxiety about mathematics, their perceptions of the appropriateness of the study of mathematics, and their opinions of the subject itself. Each of these, it appears, is influenced by a multitude of strategies and practices at home and in school which together produce gendered patterns of behaviour and achievement.

Confidence, anxiety and participation

As previously suggested, we have considerable evidence that, at all mathematical ability levels, greater uncertainty about their mathematical performance is expressed by girls than by boys (Joffe & Foxman 1984; Leder 1988; Thomas & Costello 1988). Furthermore, boys over-rate their performance in mathematics in relation to actual results while girls under-rate their performance (Joffe & Foxman 1984; Mura 1987). These same statements, however, can be and are made about academic performance generally (for example, see White 1986, about secondary students' English essays), and about many spheres of activity—boys and men overestimate and girls and women underestimate their abilities. Even leaving aside the question of why underestimating ones' abilities should be regarded as more of a problem than overestimating them, the fact remains that girls are not under-represented in secondary education or undergraduate education in Australia—presumably underestimating one's capabilities does not necessarily imply non-participation.

None the less, there is a great deal of evidence to suggest that many children and adults lack confidence in their ability to understand mathematics and, indeed, that 'panic' is not too strong a word to describe their reactions to mathematics (Cockcroft 1982). Anecdotal evidence would have it that this is more prevalent amongst girls and women than amongst boys and men, and there is a certain common sense in the view that this is a major reason why, relative to boys, fewer girls choose to study mathematics beyond the minimum levels. Certainly, studies have suggested that girls score higher than boys on tests of mathematics anxiety (Eccles & Jacobs 1986). Whether mathematics does panic more girls and women than boys and men or whether girls and women are simply more prepared to admit it is not clear since evidence of this kind is difficult to obtain. In any case, while there is evidence of a decline in gender differences in the reporting of 'maths anxiety' (Holden 1987), it is clear that mathematics panics many children and adults and this is a major problem, not of or for girls, but of and for mathematics education.

It does appear that the subject mathematics has a marked capacity to intimidate. Buxton has argued that moral judgments are often made about students' successes and failures with mathematics.

> It happens, of course, not only in maths; but the status of this subject, the clarity of correctness or incorrectness in answers, and the terms 'right' and 'wrong' with their unfortunate moral connotations, all combine to make the situation in maths more stressful than elsewhere. So failure to get the right answers is somehow seen as *morally* wrong . . . (Buxton 1981, p. 118)

Our inherited and unexamined philosophical dogma is that mathematical truth should possess absolute certainty—mathematics is about truth, mathematics is not democratic—a claim, as Agassi (1982) points out, often used to justify inept, elitist mathematics education. Mathematics curricula often convey an entirely false sense of the absoluteness of mathematics—one right answer, true in all possible worlds, dissociated from experience and independent of culture. This presentation of mathematics may alienate more girls than boys in that different patterns of behaviour with respect to obedience and authority may lead more girls than boys to become intimidated by the 'authority' of mathematics. As a consequence, they may be more powerfully influenced by the moral overtones of perceived lack of success in mathematics.

Furthermore, teaching strategies informed by such views of mathematics are often based on rote memorisation which may be even less appropriate for girls than for boys. There is, for instance, some evidence (Carss 1980) that more girls than boys feel unhappy when forced to do mathematics which they do not understand. Such feelings of unhappiness and stress could discourage even mathematically successful girls from studying the subject. For many children the result may be acute apprehension or aversion.

It has been suggested in the United States of America that the publicity given to mathematics anxiety amongst women may be making it even more socially acceptable (that is, perceived as normal) to fear mathematics (see, for example, Buerk 1985). Clearly, parents, teachers and students are all influenced by public indications that it is normal for women and men to respond to mathematics in different ways. Eccles & Jacobs (1986), for example, on the basis of a two-year longitudinal study of 250 Michigan schoolchildren of average and above average mathematical achievement as they moved from Years 7 to 9, and of their parents and teachers, concluded:

> . . . females have slightly higher math anxiety scores, and girls' mothers rate mathematics as slightly more difficult for their children than do boys' mothers. (Eccles & Jacobs 1986, p. 373)

> . . . students' grades appear to affect subsequent mathematics performance and course-taking plans only to the extent that these grades influence both parents' confidence in their children's ability to learn mathematics and students' own beliefs, motivations, and math anxiety. Furthermore, the influence of past grades on students' beliefs appears to be mediated by the parents' interpretation of these grades. (p. 374)

> . . . in this sample of mathematically competent students, math anxiety is only

> weakly related to the students' previous performance in mathematics . . . has a stronger and more direct relation to . . . students' future plans for taking math courses than does our mathematical aptitude achievement score . . . [and] is directly and strongly influenced by social factors, in particular by mothers' beliefs about the difficulty of the subject for their children. (p. 375)
>
> The results . . . also point out parents' importance as critical socializers of sex differences in mathematical achievement and attitudes. (p. 375)
>
> . . . the gender stereotypes evident in mothers' and fathers' beliefs do not appear to be grounded in reality. (p. 376)

Certainly, other North American studies (for example, Armstrong & Price 1982; Holloway 1986; Pedersen, Elmore & Bleyer 1986) have suggested that parental perceptions of their children's capabilities in mathematics influence achievement directly and indirectly through such factors as subject choice and career interest, and that parents are more likely to hold higher aspirations for their sons than for their daughters regardless of the relative achievement and interest in mathematics of the children, are more likely to be satisfied with the previous achievement of their daughters than of their sons, and are likely to provide different explanations for high performance in mathematics by their sons and daughters, respectively. Just how far this evidence can be extrapolated to the Australian context is not clear and no recent Australian evidence is available which pertains particularly to mathematics. Australian studies of career choices by girls and boys (for example, Currie 1982) indicate, however, that we should take very seriously the impact of parental influence on children's educational and occupational choices. It also seems reasonable to suggest that we take seriously the effect of the popular and media view of the relative mathematical achievement of boys and girls on what parents perceive as their children's talents and needs.

Appropriateness of the choice of mathematics

While the perception of science as a male domain is, it appears, quite strong, it is not quite so clear that this view also holds for mathematics. Recent evidence suggests that increasingly students consider mathematics as equally appropriate and necessary for both boys and girls (Joffe & Foxman 1984), although this is more true of girls than of boys, the latter still being somewhat more likely to consider girls as less capable mathematically than boys even in the face of contrary evidence in their own classrooms (Leder 1988). Nevertheless, there does appear to be some suggestion, in Western cultures, that *real* (that is, those perceived to be most feminine) girls would not want to do mathematics and that not doing mathematics is more socially acceptable for girls than for boys.

Related to this is girls' and boys' perceptions of the usefulness of mathematics. It seems that boys and girls equally regard mathematics as useful (Kelly et al. 1986), but that, as they proceed through secondary schooling, girls become less convinced that mathematics is personally useful. In the United Kingdom, Russell (1984) reported that neither female nor male students considered that the content of their O-level mathematics course was particularly useful, but the boys saw mathematics generally as somewhat more useful than did the girls because they considered the qualification itself as more useful for their future occupations. Russell's study suggested that the boys chose mathematics at O-level in spite of their perceptions of the irrelevance of the content, partly because it is needed for other things, partly because it was the expected thing to do and partly for its status with teachers and peers.

Both girls and boys recognise that mathematics is used as a filter, as this conversation of a group of Western Australian children in Year 8 illustrates:

[Int]: Do you need maths?
Boy 2: Yeah. Yeah. But . . .
Girl 1: You need basic maths but you don't really need to know this. ['This' is algebra.]
[Int]: You don't need to know this?
Girl 2: No I mean. You just need the basic maths.
[Int]: What's basic maths?
Girl 1: You know adding fractions and, you know, timesing and everything. Adding and subtracting. 'Cause that's all really you meet up with. The rest you can do on the calculator.
[Int]: Saul do you reckon you'll need this sort of maths?
Saul: You only do it because they use it to see if you understand and if your brain knows how to do it.
[Int]: And do you think it's important that your brain knows how to do it? Will you need to know how to do it for your later life?
Saul: Yeah, you need to know how to do it. That's why we do it. To make sure we know how to do it. (The other students laugh.)
[Int]: That sounds great. But where in your later [life], do you need to do pyramids?
Saul: Oh they just use it to see how fast you can work out things and that, how you catch on, and you understand them.
[Int]: Oh, how you catch on?
Saul: And you can work them out.
Girl 2: They just want to see if you don't understand it. (Holland 1987, p. 139)

The different participation rates of boys and girls in mathematics indicate quite clearly that its use as a filter for a wide range of occupations is insufficient at present to convince girls to participate to the same extent as boys. Interestingly, both adults and children who appear to think mathematics equally important for boys and girls none the less regard it as more important for adult men than for adult women (Handel 1986). According to Handel, girls who believe mathematics to be less relevent for adult women than for adult men are, not surprisingly, less likely to participate in the highest levels of mathematics.

Rather than limited post-school options being a consequence of limited participation in mathematics, the reverse may be true (Armstrong & Price 1982; Meece, Eccles & Parsons 1982). While it is often suggested that girls underparticipate and underachieve in science and technology-related courses and occupations because these are more quantitative and because girls lack the interest and background in mathematics, it is arguable that girls decide not to study higher levels of mathematics because they do not want to study science and technology and regard these as the main areas in which mathematics is applied. Girls intending to work in areas which traditionally have not, or which they believe have not, demanded high levels of mathematics may consider it a waste of time to study the subject.

An interesting study, published in the United States of America in 1985, offers more insight into this matter. The study reported that mathematically able females are somewhat more likely than their male peers to have multiple talents and that whether or not mathematically talented girls chose to study mathematics further was related less to their conceptions of their mathematical abilities and more to their conceptions of their range of abilities (Hollinger 1985). That is, those girls who perceived their talents to be mostly in the mathematics area chose mathematics or science-related careers to the same extent as did their male peers, but those girls who perceived themselves to have a range of options were less likely to chose mathematics. This seemingly obvious factor is overlooked in many egocentric analyses of girls' lower participation in mathematics which appear to assume that if you can do mathematics and know you can, and know of the 'doors it opens', then you will so choose. These girls, clearly, are deciding that other doors are open also and that they prefer to go where some of them lead.

The pressures on 'bright' girls regarding achievement and participation in mathematics often appear different from those on 'bright' boys. Leder (1977) studied the relationship between results on a mathematics test and long-term occupational aspirations of Victorian boys and girls in Years 10 and 11. She found that 'motive to avoid success' was strongest in those girls who aimed for male-dominated, high-status occupations, 'unusual' for women. Leder commented that

> an increasing realization that attainment of an ambitious goal may be a mixed blessing and may have negative personal consequences may well lead to a lowering of personal goals. Alternately, the growing anxiety about the consequences of attaining an ambitious goal may act as an impediment on performance ... By contrast, those girls who have decided to opt for a more traditional female career, irrespective of their mathematics performance, no longer seem subject to such conflict. (Leder 1977, pp. 186–7)

There has been considerable criticism of constructs such as 'fear of success' and 'motive to avoid success' (Sassen 1980), but it is important to note that what Leder is referring to here are the real difficulties likely to be faced by girls who choose mathematics (or other male-dominated subjects and careers); these difficulties are not simply the overactive imaginings of adolescence.

Many of the post-secondary courses and occupations (apart from teaching) which actually require high levels of mathematics are prestigious and relatively highly paid and access to such courses and occupations should not be sexist (or for that matter classist or racist). On the other hand, many of the more prestigious occupations are limited in number and the associated career structures are such that they may inhibit involvement by many girls and women. Convincing more girls to study mathematics is unlikely to reduce the institutionalised and sexist practices which require that job entry and progress in many professions must occur during the ten major childbearing years. More flexible structures are possible but many of the girls making their decisions about future occupations have reason to be sceptical about the speed with which change will occur. Governments allocate funds to increase the participation of girls in mathematics and science at the same time as they refuse to recognise adequate childcare facilities as a right. Industries clamour for more qualified mathematicians, scientists and technologists but neglect to deal with the industrial issues related to part-time work and disrupted periods of work caused by childbearing. As many women know only too well, the 'professional culture' of many higher status occupations assumes that the home responsibilities of the professional 'person' can take second place to the work responsibilities and that such a person has a family support system to provide housekeeping and childcare. At present, such occupations are simply not equally available to those who need reasonable and regular working hours and who cannot readily travel. Such problems exist for women entering the workforce generally but certainly are more intractable in the occupations identified as being part of the male domain. On the other hand, it need not be the case that parenting and employment are in conflict; it is possible to set up community and employment structures to enable men and women to accommodate both sets of responsibilities.

These issues are not only of relevance to girls who are middle-class and mathematically more able, of course; rather, they are part of a whole constellation of educational, occupational and family structures which constrain girls' choices in ways which the girls themselves often cannot even begin to understand and which are, therefore, even more powerfully restricting. In particular, the absence of a significant proportion of women in these male-dominated professions results in a climate in which it is generally less accepted for girls to enter the related domains. Even further, the understanding that for women their families must come first and their jobs a poor second must influence girls' aspirations and educational and occupational decisions. Interestingly, in England, while 'Anglo' boys achieve more highly than 'Anglo' girls at 16+ examinations in mathematics, West Indian girls achieve more highly than both 'Anglo' boys and West Indian boys (Driver 1980). West Indian women in the United Kingdom are often the family provider, and education for women is regarded as the most likely avenue for better jobs. These girls regard having a good job as an important part of being a good wife and mother. As already indicated, evidence of this kind is also available about sub-cultural groups in the United States of America.

Many girls do believe that there is a conflict for women between family life and careers. Their educational choices are influenced by their expectations of how they, as adults, will allocate their time between the labour force and home. Some of these expectations are realistic and others not so; either way, they tend to become self-perpetuating. In many cases, girls are making choices in those periods of their lives when they are most likely to be influenced by peer group and pop culture, and by romantic thoughts of boyfriends, marriage and babies. Therefore, their aspirations are less likely to be directed at careers which do not fit into the feminine stereotype.

Currently, for most students, the major rewards from continued participation in mathematics are extrinsic through credentials and prospective occupations. It seems that the influence of these extrinsic rewards is, for a range of reasons, less powerful on most Australian girls than on boys and we define this as a 'problem'. But the problem needs to be turned on its head. The real problem for mathematics education is that the majority of girls and boys do not gain intrinsic rewards from mathematics (Cockcroft 1982).

Opinion of the subject mathematics

Girls and women are generally considered to be more interested in and skilled to deal with social and interpersonal issues than are boys and men. The inference is then drawn that girls and women are relatively less suited to mathematics because it is a 'purely intellectual', 'more abstract' pursuit. This is a particularly fascinating reading of the present situation where girls less often than boys remain in the highest levels of mathematics, because it in no way matches the evidence we have of the areas in which girls (logic) and boys (measurement, geometry) excel in mathematics. It is, however, perhaps not a surprising reading because it enables one to do nothing (unless or until the nation finds it needs the girls, as is the case presently). Female culture and the culture of mathematics differ in fundamental ways—so be it!

Nevertheless, there is reason to believe that girls are influenced in their choice of subjects by their perceptions of the subjects' social relevance and capacity to improve the quality of life (Soberg & Imsen 1988). Interestingly, one report from a large technology competition in which students constructed their own problems indicated that girls said that the problems they choose involved using technology to serve a social purpose, such as designing for the handicapped, while boys' explanations of their choices related to the technology itself (Harding 1987/8).

It also seems that girls' involvement in mathematics is more inhibited than boys' by the perception of mathematics as a subject that is culture-free and dissociated from experience. It has been argued that girls are more likely to prefer more context-bounded tasks and are less prepared than boys to accept the version of mathematics as 'disembedded thought'. But mathematics *as* disembedded thought is a particular version of mathematics, or perhaps I should say a part, but not the whole, of what mathematics is. Brown (1984) has argued that there is not just one mathematics, that 'what has constituted legitimate thinking in the discipline has changed considerably over time' (pp. 13–14) and that it is a masculine interpretation of mathematics that produces a curriculum which is '"de-peopled" in that contexts and concepts are for the most part presented ahistorically and unproblematically' (p. 12). Such a mathematics is about absolutes, about 'taken for granted' reality upon which students are to operate, about similarities rather than differences, about problem solving rather than problem posing. A mathematics influenced by 'female' perspectives might, he argues, emphasise context boundedness, differences between situations and problems, and 'people connectedness'. It is important to emphasise that what Brown has described as a masculine construction of mathematics is not essentially male but reflects a particular construction of masculinity also inappropriate for a great many boys. School curricula always represent a selection from a much wider universe of knowledge, and mathematics is no different. In particular, the content and pedagogy of a great deal of school mathematics excludes the majority of children.

Earlier it was suggested that gender differences in liking or disliking mathematics no longer appear in the primary and secondary years. None the less, the liking or disliking of mathematics during these early years does influence the choice of mathematics once it becomes optional, and it appears that the effect is greater for girls than for boys (Russell 1984). The success of mathematics in attracting middle-class males is more apparent than real, being more closely related to its prestige and its status as a 'critical filter' between school and employment than to any fundamental mathematical qualities of the courses offered. As Russell (1984) suggests, these boys are a 'captive audience' because they would study upper secondary mathematics almost regardless of the curriculum.

One way of interpreting the problem of girls and mathematics is that those who feel they have a choice, and feel able to decide not to study mathematics, do so decide! But, of course, it is not so simple. From the preceding material it should be clear that in a range of ways girls are less 'free to choose' to participate in mathematics and science than are boys, and boys are correspondingly less 'free to choose' not to participate. This happens through implicit and explicit assumptions and messages about male and female intellectuality, inclinations and needs, and about male and female futures. Many teachers do believe that girls, through ability or inclination, are less capable than boys of succeeding in the empowering male domains of mathematics and science. Others believe that, even if girls can succeed in these fields, they need not. Still others share the view of some students and their parents that there is little point in young women trying to change the status quo, that they will be happier, in the long term, if they 'make choices' consistent with their future responsibilities as childrearers and homemakers. Such teachers may consider it unfair but inevitable. Schools, they may argue, do not change society, but rather reflect it. And thus gendered educational experiences and outcomes are justified. As Buswell (1984) has suggested of United Kingdom schools, while in theory schools offer equality, they also engage in a process of 'cooling out' aspirations in girls in preparation for a long-term future of inequality. The pressures on boys and girls may, of course, be equally strong, but the effect in terms of post-school options is quite asymmetrical.

Chapter 5
Which way forward?

'Would you tell me, please, which way I ought to go from here?'
'That depends a good deal on where you want to get to,' said the Cat.
Lewis Carroll, *Alice's Adventures in Wonderland* (1865)

Comparatively recently, girls and women were formally excluded from the study of higher levels of mathematics (Clements 1979). The formal barriers no longer exist — to the contrary, as I remarked in my introduction to this monograph, governments presently allocate funds to the task of increasing the participation of girls in mathematics and, consequently, they hope, in science. It is surely no accident that increasing the number of women in mathematics and science (always as a pair, lest there be a slippage of women from mathematics to commerce and economics) becomes a priority at a time when 'bright' boys are choosing to leave these areas in droves (see, for example, *Science and Technology Education in Western Australia: A Matter for Concern* 1987). As *Rosie the Riveter* was told, girls are now being told that their nation needs them; women are to undertake the (relatively poorly paid) basic mathematics and science work while 'their men' get on with the real (relatively highly paid) work of commerce and economics. And yet, cynicism aside, we must ask whether we do want to increase the participation of girls and women in mathematics and, if so, why.

Why more mathematics?

Mathematics is used as a critical filter between school and the workplace (at times mediated by post-secondary education) and many regard this as sufficient reason to encourage girls to take more high-level mathematics. But we should question efforts to convince girls to study mathematics on the basis of improved job prospects. Even if we ignore for the moment the argument presented earlier that it is structurally more difficult for women to succeed in many of the male-dominated employment domains, it is still not clear that in the future the majority of jobs will be mathematics and science related. Consider the following extracts from the Quality of Education Review Committee Report:

> For the future, the types of employment most likely to grow appear to be those in installation, maintenance and repair, information processing, administration, clerical and other office activities, and personal services, both public and private. They are . . . [not] clearly associated with particular formal educational qualifications. At the same time they tend to involve high degrees of client contact and interpersonal skills . . . (QERC 1985, p. 57)
>
> A much greater proportion of young Australians will have to complete full secondary school and receive a broad general education as a foundation on which to

build the occupational skills necessary to operate in a labour market in a state of flux and in which the capacity to deal with people will become a growing requirement. (pp. 59-60)

According to the QERC report, then, future jobs are likely to require those very interpersonal skills at which girls and women are often regarded as excelling. This, together with their general high level of educational success, should prepare them well for the areas in which future jobs are likely to lie.

But, again, it is not quite so simple. Girls (and working-class children generally) who chose not to study school mathematics *do* restrict their access to many tertiary courses and a variety of occupations whose entrance requirements include certain levels of mathematics. Often they are restricted because they do not have access to some particular aspects of mathematics, for example, calculus, but, equally often, mathematics is used simply as a filter. Faculties in several Australian universities, for example, require a Year 12 mathematics course at the highest level but indicate that any branch of mathematics will do. Clearly, all that is required is evidence that the student can do mathematics at a particular level; as one of the Year 8 students quoted earlier remarked, '. . . they just use it to see . . . how you catch on, and you understand them' (Holland 1987, p. 139). Often, the preparatory or prerequisite nature of certain school subjects for tertiary preparation is confused with the selective function.

Mathematical prerequisites for courses and job entry are rarely justified explicitly in terms of the concepts to be studied or the tasks to be undertaken (Bannister 1987). Recently, a psychologist asked me to assist him to check whether the existing mathematics test for applicant prison officers was consistent with the Year 10 mathematics curriculum. Since the test used only imperial units, I could have answered the question fairly quickly. Instead, I suggested that, rather than preparing a new set of textbook exercises based on the school curriculum, the prisons department identify the kinds of problems and situations (mathematical or otherwise) a prison officer would be likely to confront, and design their battery of tests around those. Not too surprisingly, my suggestion met considerable resistance. My offhand comment that, since applicant prison officers must be several years removed from Year 10, there would always be a time lag in using the school curriculum as a basis for selection was received somewhat more warmly than my argument that entrance tests should relate to the nature of the future position. I was told, essentially, that what I suggested was neither necessary nor desirable. 'We just use the maths test to pick the most intelligent people. You do think [rather accusingly] that prison officers should be intelligent, don't you, and that more intelligent ones must be better than less intelligent ones?'

Mathematics is not used as a selection device simply because it is useful, but rather the reverse. It is regarded as useful largely because it is a powerful selector, taking the role of the now widely discredited IQ tests but without receiving the same criticisms (after all, mathematics is widely regarded as culture-free). Consider, for example, the case of ASAT. This test 'is designed to be independent of any particular school curriculum . . . [and to] test a student's capacity for intellectual pursuits at the tertiary level' (Adams 1986, p. 267). It is used in Western Australia to moderate external examination results in different subjects, and 10% of a student's ASAT score is added to the tertiary entrance aggregate. In Queensland and the ACT it is used to moderate the internal assessments of schools and subjects and in the ACT the score has appeared on students' certificates and has been used in the selection processes of the Australian National University. In each case, although to varying degrees, differences between the average scores of boys and girls are found in favour of the boys. These gender differences are not reflected in school-based or examination results, or *in the later success of the students in tertiary education*. Adams analysed the ACT results on this test and concluded that

> ... no significant direct effect of sex on performance on ASAT was found. Further, sex differences in performance were found to be the result of females having a higher verbal ability as measured by English assessments, males studying more mathematics, and males having greater confidence in success on ASAT. (Adams 1986, p. 273)

Even though the test is supposed to be free of any effect of particular course taking, Adams contends that the fact that the results favour boys who study more mathematics cannot be construed as a gender bias (or for that matter a bias against any group of students who choose less mathematics) on the basis of the rather staggering assertion, in this context, that 'it is not unreasonable to expect that it is those students of greater quantitative aptitude who would study more mathematics' (Adams 1986, p. 274). That this is not so is made palpably clear by the examination results in Year 12 mathematics in Western Australia which show quite clearly that many mathematically very able girls do not take 'double maths' and many boys who do so have clearly made a quite inappropriate choice (Parker 1984).

An irony of this, of course, is that were this hold (directly through imposed prerequisites and indirectly through, for example, ASAT) of mathematics on post-secondary options to be broken we would be likely to see a drop in the numbers of students (particulary boys) enrolled in the highest level Year 12 courses (see Widdup 1981; and also the Western Australian experience in 1988 under new, more liberal tertiary admissions structures). None the less, maintaining participation rates in mathematics by erecting false educational and occupational barriers can hardly be justified from the perspective of either the mathematically more able girls who do not choose to undertake the demanded level of mathematics study or the mathematically less able boys who do but, on the basis of their results, should not. Certainly, as Parker suggests, it is doubtful that many of the latter students benefited from two years' study of a double mathematics course from which they apparently learned staggeringly little.

Thus school mathematics is a filter, and it is perhaps not surprising that currently it filters out much the same people as did IQ tests or that it manages to sort people readily on the basis of gender, class and race. It is the carefully fostered belief that all people who can do competitive forms of mathematics are more intelligent than all those who cannot, together with the practices associated with this view, which should be the focus of attention of those who wish to assist girls and others who are excluded from many occupations in the name of mathematics. In my view, mathematics is too good to be used for such dishonest ends.

Having argued that convincing girls to study mathematics because of hypothetical job prospects seems problematical, I would now like to suggest that, indeed, the broad range of girls and boys *should* have increased access to mathematics. Mathematics, both its bodies of knowledge and its ways of knowing, represents a significant aspect of human activity. It has intrinsic merit through its potential to enrich our understanding of physical, social and intellectual apects of our lives and through the pleasure and empowerment it can provide. That merit is distorted, however, when access is limited to so few. It is hard to escape the notion that its exclusive status is maintained because it serves the interests of a privileged minority.

For those who do not continue in school beyond the school leaving age, the problems of limited access to the empowering forms of mathematics are perhaps most urgent because mathematics does have another powerful role in our community not unrelated to its selection role. That is, mathematics is widely used to intimidate those to whom access is limited. Koblitz (1981), in an excellent article entitled 'Mathematics as propaganda', has described many examples, often taken from 'respectable' publications, of the use of mathematics in highly ambiguous ways to produce mystification and an impression of precision and profundity. Arguments that would be ridiculed if explained in everyday language are accepted when presented 'mathematically' — they are then regarded as scientific because they involve 'hard' data. Invoking numbers, statistics and formulae can be more persuasive than quoting well-known authorities, and in the presence of mathematics many of us suspend our disbelief. For example, why *don't* more of us argue about the mathematically defined tertiary admissions strategies that so many of us distrust? Perhaps we are intimidated by the apparent authority and sureness of the mathematics — after all, 'numbers don't lie'.

Many adults, as children, were taught in mathematics classrooms to trust implicitly processes they did not understand. Many are now intimidated by mathematics; they feel inadequate in its presence. For this reason, girls and women (and other mathematically disadvantaged groups) must gain access to the kind of mathematics that enables them to resist such attempts to intimidate. Encouraging more girls to study more mathematics may be necessary but it will not be sufficient — we must ensure that the mathematics they study is truly empowering. Thus, we need to address seriously in mathematics education the issue of 'whose knowledge' has status, of what mathematics we teach and of how we teach it. As Yates has argued,

> the work on sexism has made clear that being fair to girls is likely to imply more than simply not overtly discriminating against them, and more than simply being nice to girls and adding in examples of women in the curriculum. It raises questions about whether the lines of past practice might need some changes of direction and about competing interests that might need some adjudicating. (Yates 1988, p. 43)

What might a mathematics curriculum and classroom which was more empowering for girls look like? We do have some indications from the explanations offered earlier for girls' lower participation and achievement in higher level mathematics. These were varied and not all equally accessible to the influence of schools, but broad implications for curriculum content and pedagogy can be drawn from them.

Towards an empowering mathematics curriculum

Webber, in an article entitled 'Maths as a subversive activity', has suggested that mathematics for empowerment involves the development of the capacity to grapple with problems until a critical understanding is reached. Further,

> it means learning to create as well as solve problems, to ask questions, to gather and extract useful information, to criticise assumptions, and to use numbers to support or refute opinions. It means learning to cooperate and share ideas, and to place the mathematical component of a problem in a meaningful context. It means owning all aspects of the learning experience. (Webber 1987/8, p. 8)

It is often suggested that the 'newer' curriculum developments will ensure that the mathematical experiences of girls (and currently excluded boys) will become more empowering, and certainly this is a reasonable assertion at face value. Newer teaching strategies incorporate a range of ways of 'seeing' the subject and encourage collaboration, the pooling of ideas and the sharing of control. They also place a greater emphasis on co-operative rather than competitive learning, while attempting to help students develop a sense of control over their own learning. Learning experiences are recommended which include extended investigations and mathematics in real-world contexts. All of these *do* have the potential to make mathematics learning more rewarding and productive for the majority of students, including girls. Similarly, pedagogy which accepts error as an essential part of learning and assists students to deal with error and conflict, which encourages students to share partial solutions and different interpretations of problems, which provides time for, and acknowledges the importance of, reflection and clarification of ideas, and which places an increased emphasis upon the processes of mathematics rather than only the content, may provide girls with a changed view of the demands of mathematical thinking. But it is unlikely that these alone will change the mathematical experiences of girls unless the curriculum explicitly addresses their needs.

Towards a mathematics curriculum inclusive of girls

Clearly, a curriculum is needed which is non-sexist (and also non-classist and non-racist) in content, language, illustrations, examples and representations of males and females and which acknowledges and incorporates women's contributions both to mathematics and to society. Equally, we would expect of such a curriculum that it be rigorous and intellectually demanding, neither providing nor risking accusations of providing what are often derisively called 'soft options'. The latter, clearly, would disservice girls directly by limiting their access to more empowering forms of mathematics and indirectly through the inferences made about the apparent need for such options. Brown (1984, p. 14) has suggested that 'what may be called for is an ever more intellectually demanding curriculum, but one in which mathematics is embedded in a web of concerns that are more "real world" oriented than any of us have begun to imagine'. Thus we would expect to embed mathematics in real-world social concerns and in people-oriented contexts; presenting it as making 'humansense', non-arbitrary, non-absolute and also fallible; and presenting a social–historical perspective to help students become aware of the 'person-made' quality of mathematics. Further, we would emphasise the aesthetic and cultural values of mathematics as well as the instrumental values, rejecting narrowly instrumental interpretations of mathematics while demonstrating that its instrumental uses are very broad and not restricted to a few scientific careers.

A mathematics curriculum such as has been described in the preceding two paragraphs *is* possible. It is the dream of many who involve themselves in the mathematics education of both boys and girls. I believe it is even becoming more likely. But will it be enough? Probably it will not, unless the very beliefs and practices which underpin the learning and teaching of mathematics also become non-sexist. We know from the work of Walkerdine and others that both our perceptions of what 'natural' mathematical thinking looks like and our conceptions of learning as the drawing out of natural behaviour are very sexist indeed. Little of substance will change in mathematics classrooms for girls (and indeed for a great many boys) until we become more sensitive firstly, to the ways in which different, often stereotyped, readings of success and failure, and their associated classroom practices, label students as good or bad at mathematics and secondly, to the struggle some girls have in handling the apparently conflicting demands of 'femininity' and what they regard as 'masculine' forms of intellectuality.

Finally, an inclusive mathematics curriculum for girls would address the issue of gender formation in schools explicitly, enabling both boys and girls to identify the processes by which gendered patterns of achievement and participation in mathematics is produced and naturalised, and to recognise the effect of these processes on their futures. As Blackburn (1982) has suggested, the role of education with respect to gender formation is to help young people 'to understand and reflect on what is, on how it came to be that way and on what they want to do about it, both at the level of their own lives and in relation to social action' (p. 10). Some would argue that this is not the responsibility of mathematics teachers, but anything short of a critical understanding of the role mathematics currently plays in constructing privilege and the role it could play in deconstructing myths can never be truly empowering.

Conclusion

The purpose of this monograph was to discuss issues related to gender and mathematics education. It began by pointing to the evolution in our definition of the 'problem of girls and mathematics' — once girls couldn't do mathematics as well as boys, then they probably could but didn't, and now they simply won't. As the monograph has shown, neither achievement differences nor participation differences are as great as is often assumed, although there certainly are differences in both — particularly at the highest levels of achievement. A range of explanations is offered in the literature for those gender differences which do exist. As has become clear, the problems are complex and unlikely to be amenable to simple solutions. In particular, while many of the problems of girls' mathematics education are but symptomatic of problems in the mathematics education of the majority of children, others are quite clearly bound up with gender-related assumptions which inform all educational practice.

One of the major difficulties we face is the implicit assumption that 'natural' differences in aptitudes, inclinations and futures cause gender differences in mathematical experiences. Another is the assumption that the problem of girls and mathematics is one of and with girls. In this monograph, I have argued that we need to turn the problem on its head and ask what needs to happen in classrooms to increase girls' real access to the empowering forms of mathematics. Clearly, that is an ambitious project, and only some elements of it have been suggested in the last section.

References

Adams, R. (1986), 'Some contributions to sex differences in scholastic aptitude scores', *Studies in Educational Evaluation* **12**(3), 267–74.

Adams, C., & Walkerdine, V. (1986), *Investigating Gender in the Primary School: Activity-based Inset Materials for Primary Teachers*, Inner London Education Authority, London.

Agassi, J. (1982), 'Mathematics education as training for freedom,' *For the Learning of Mathematics* **4**(1), 28–32.

Armstrong, J. (1980), *Achievement and Participation of Women in Mathematics: An Overview*, Education Commission of the States, Denver.

Armstrong, J., & Price, R. (1982), 'Correlates and predictors of women's mathematics participation', *Journal for Research in Mathematics Education* **13**(2), 99–109.

Atweh, W. (1980), *Sexist Attitudes in Mathematics Education*, Mathematics Department, Kelvin Grove College of Advanced Education, Brisbane.

Bannister, H. (1987), 'Gender and tertiary selection', Research Paper no. 2, Participation and Equity Program, Ministry of Education, Melbourne.

Barnes, M. (1987/8), 'The power of calculus', *Education Links* (32), 25–7.

Benbow, C., & Stanley, J. (1980), 'Sex differences in mathematical ability: Fact or artifact?', *Science* **210**, 1262–4.

Blackburn, J. (1982), 'Some dilemmas in non-sexist education', *The Secondary Teacher* **3**(1), 10–11.

Bourke, S., & Stacey, K. (1988), 'Assessing problem solving in mathematics: Some variables related to student performance', *Australian Educational Researcher* **15**(1), 73–83.

Brandon, P., Newton, B., & Hammond, D. (1985), The superiority of girls over boys in mathematics achievement in Hawaii, Paper presented at the 69th Annual Meeting of the American Educational Research Association, Chicago, April.

Brown, S. (1984), 'The logic of problem generation: From morality and solving to de-posing and rebellion', *For the Learning of Mathematics* **4**(1), 9–20.

Buerk, D. (1985), 'The voices of women making meaning in mathematics', *Journal of Education* **167**(3), 59–70.

Buswell, C. (1984), 'Sponsoring and stereotyping in a working-class English secondary school', in S. Acker, J. Megarry, S. Nisbet & E. Hoyle (eds), *World Yearbook of Education 1984: Women and Education*, Kogan Page, London.

Buxton, L. (1981), *Do You Panic About Maths?: Coping with Maths Anxiety,* Heinemann Educational Books, London.

Campbell, J., & Connolly, C. (1984), Impact of ethnicity on maths and science among the gifted, Paper presented at the 68th Annual Meeting of the American Educational Research Association, New Orleans, April.

Carss, M. (1980), 'Girls, mathematics and language: Some observations from classrooms', *Improving Maths for Girls,* Report of a conference held at Raywood Inservice Centre, Adelaide, May/June, Education Research and Development Committee.

Carss, M. (1982), 'The language of mathematics', in R. Kennedy (ed.), *Expanding the Options—Girls, Mathematics and Employment,* Social Development Unit, New South Wales Department of Education, Sydney.

Chipman, S. (1981), Letter to the editor, *Science* **212**, 114-6.

Clarkson, P., & Leder, G. (1984), 'Causal attributions for success and failure in mathematics: A cross cultural perspective', *Educational Studies in Mathematics* **15**(4), 413-22.

Clements, M. (1979), 'Sex differences in mathematical performance: An historical perspective', *Educational Studies in Mathematics* **10**, 305-22.

Clements, M., & Wattanawaha, N. (1977), 'Sex and age-within-grade differences in mathematical achievement of Victorian children', *Research in Mathematics Education in Australia,* **2**, 9-36.

Cockcroft, W. (1982), *Mathematics Counts,* HMSO, London.

Currie, J. (1982), 'The sex factor and occupational choice', *Australian and New Zealand Journal of Sociology* **18**(2), 180-95.

Day, N. (1987), 'Research report: Factors associated with student choice of mathematics, science and technology in Years 11 and 12 in Victoria', in *Improving the Science, Mathematics and Technology Education Base in Australia,* Report of the Working Group on Education for Science and Technology to the Victorian Government, Melbourne.

Dekkers, J., De Laeter, J., & Malone, J. (1986), *Upper Secondary School Science and Mathematics Enrolment Patterns in Australia, 1970-1985,* Curtin University of Technology, Bentley, WA.

Donaldson, M. (1978), *Children's Minds,* Fontana, Glasgow.

Driver, G. (1980), 'How West Indians do better at school (especially the girls)', *New Society* **51**(902), 111-14.

Dweck, C. (1986), 'Motivational processes affecting learning', *American Psychologist* **41**(10), 1040-8.

Easley, J., & Easley, E. (1982), *Math can be Natural: Kitamaeno Priorities introduced to American Teachers,* Committee on Culture and Cognition, University of Illinois, Urbana, IL.

Eccles, J., & Jacobs. J. (1986), 'Social forces shape math attitudes and performance', *Signs: Journal of Women in Culture and Society* **11**(2), 367-80.

Education Gazette, South Australia (1979), **7**(30), 27 September.

Edwards, J. (1985), 'Boys and girls in the Australian Mathematics Competition', *Mathematics in School* **14**(5), 5-7.

Egelman, E., Alper, J., Leibowitz, L., Beckwith, J., Levine, R., & Leeds, A. (1981), Letter to the editor, *Science* **212**, 116.

Ethington, C., & Wolfle, L. (1984), 'Sex differences in a causal model of mathematics achievement', *Journal for Research in Mathematics Education* **15**(5), 361-7.

Evans, T. (1982), 'Being and becoming: Teachers' perceptions of sex-roles and actions towards their male and female pupils', *British Journal of Sociology of Education* **3**(2), 127–43.

Fennema, E. (1981), Letter to the editor, *Science* **222**, 116.

Fennema, E., & Peterson, P. (1985), 'Autonomous learning behavior: A possible explanation of sex-related differences in mathematics', in E. Fennema (ed.), 'Explaining sex-related differences in mathematics: Theoretical models', *Educational Studies in Mathematics* **16**(3), 303–20.

Fennema, E., & Peterson, P. (1986), 'Teacher–student interactions and sex–related differences in learning mathematics', *Teaching and Teacher Education* **2**(1), 19–42.

Fennema, E., & Tartre, L. (1985), 'The use of spatial visualization in mathematics by girls and boys,' *Journal for Research in Mathematics Education* **16**(3), 184–206.

Galbraith, P. (1986), 'The use of mathematical strategies: Factors and features affecting performance', *Educational Studies in Mathematics* **17**(4), 413–41.

Gelman, D., & Carey, J. (1981), 'Sex research – a history of bias', *Newsweek*, 1 June, 47.

Grimison, L. (1982), 'An analysis of mathematics textbooks for sexism', in R. Kennedy (ed.), *Expanding the Options – Girls, Mathematics and Employment*, Social Development Unit, New South Wales Department of Education, Sydney.

Hallinan, M., & Sorenson, B. (1987), 'Ability grouping and sex differences in mathematics', *Sociology of Education* **60**(2), 63–72.

Handel, R. (1986), Achievement attitudes in mathematics and science: Relationship between self-perceptions, aspirations, and extra-curricular activities, Paper presented at the 70th Annual Meeting of the American Educational Research Association, San Francisco, April.

Hanna, G., & Kuendiger, E. (1986), *Differences in Mathematical Achievement Levels and in Attitudes for Girls and Boys in Twenty Countries*, Ontario Institute for Studies in Education, Toronto.

Harding, J. (1987/8), 'Filtered in or opting out?', *Education Links* (32), 12–14.

Harris, L. (1978), 'Sex differences in spatial ability: Possible environmental, genetic and neurological factors', in M. Kinsbourne (ed.), *Asymmetrical Function of the Brain*, Cambridge University Press, New York.

Holden, C. (1987), 'Female math anxiety on the wane', *Science* **236**, 660.

Holland, P. (1987), Formalization in children's initial learning of algebra, MEd thesis, Murdoch University, WA.

Hollinger, C. (1985), 'Self-perceptions of ability of mathematically talented female adolescents', *Psychology of Women Quarterly* **9**(3), 323–6.

Holloway, S. (1986), 'The relationship of mothers' beliefs to children's mathematics achievement: Some effects of sex differences', *Merrill-Palmer Quarterly* **32**(3), 231–50.

Horner, M. (1972), 'Towards an understanding of achievement-related conflicts in women', *Journal of Social Issues* **28**, 157–75.

Hunt, G. (1985), 'Math anxiety – where do we go from here?', *Focus on Learning Problems in Mathematics* **7**(2), 29–40.

Husen, T. (ed.) (1967), *International Study of Achievement in Mathematics*, 2 vols, Wiley, New York.

Hyde, J. (1981), 'How large are cognitive gender differences?', *American Psychologist* **36**(8), 892–901.

Joffe, L., & Foxman, D. (1984), 'Assessing mathematics 5: Attitudes and sex differences', *Mathematics in Schools* **13**(4), 22–6.

Jones, W. (1988), 'Entry to engineering, technology, and science in higher education', Occasional Paper no. 20, Australian Council for Educational Research, Hawthorn, Vic.

Keeves, J., & Bourke, S. (1976), *Australian Studies in School Peformance*, vol. 1, Education Research and Development Committee, AGPS, Melbourne.

Keeves, J., & Mason, J. (1980), 'Sex differences in attitudes towards achievement and participation in mathematics in school', *Improving Maths for Girls*, Report of a conference held at Raywood Inservice Centre, Adelaide, May/June, Education Research and Development Committee.

Kelly, A., Alexander, J., Azam, U., Bretherton, C., Burgess, G., Dorney, A., Gold, J., Leaky, C., Sharpley, A., & Spandley, L. (1986), 'Gender roles at home and school', in L. Burton (ed.), *Girls into Maths can go*, Holt, Rinehart & Winston, London.

Kenway, J., & Willis, S. (1986), 'Feminist single-sex strategies: Some theoretical flaws and practical fallacies', *Discourse* **7**(1), 1–30.

Kissane, B. (1986), 'Selection of mathematically talented students', *Educational Studies in Mathematics* **17**(3), 221–41.

Koblitz, N. (1981), 'Mathematics as propaganda', in L. A. Steen (ed.), *Mathematics Tomorrow*, Springer-Verlag, New York.

Leder, G. (1974), 'Sex differences in mathematics: Problem appeal as a function of problem context', *Journal of Educational Research* **67**(8), 351–3.

Leder, G. (1976), 'Contextual setting and mathematics performance', *Australian Mathematics Teacher* **32**(4), 119–27 and **32**(5), 165–73.

Leder, G. (1977), 'Mathematics performance and future occupation: Are they related?', *Research in Mathematics Education in Australia* **1**, 179–88.

Leder, G. (1980a), 'Bright girls, mathematics and fear of success', *Educational Studies in Mathematics* **11**(4), 411–22.

Leder, G. (1980b), 'Sex differences in achievement and participation in mathematics: A selective review', *Improving Maths for Girls*, Report of a conference held at Raywood Inservice Centre, Adelaide, May/June, Education Research and Development Committee.

Leder, G. (1982), 'Mathematics achievement and fear of success', *Journal for Research in Mathematics Education* **13**(2), 124–35.

Leder, G. (1984), 'Sex differences in attributions of success and failure', *Psychological Reports* **54**(1), 57–8.

Leder, G. (1985), 'Sex-related differences in mathematics: An overview', *Educational Studies in Mathematics* **16**(3), 304–9.

Leder, G. (1986), Teacher–student interactions—a hidden message, Paper presented at the Annual Conference of the Australian Association for Research in Education, Melbourne, November.

Leder, G. (1987), 'Teacher student interactions: A case study, *Educational Studies in Mathematics* **18**(3), 255–71.

Leder, G. (1988), 'Teacher-student interaction: The mathematics classroom', *Unicorn* **14**(2), 107–11.

Leinhardt, G., Seewald, A., & Engel, M. (1979), 'Learning what's taught: Sex differences in instruction', *Journal of Educational Psychology* **71**(4), 432–9.

Lovejoy, F., & Barboza, E. (1984), 'Feminine mathematics anxiety: A culture-specific phenomenon', in R. Burns & B. Sheehan (eds), *Women and Education*, La Trobe University, Bundoora, Vic.

Maccoby, E., & Jacklin, C. (1974), *The Psychology of Sex Differences*, Stanford University Press, Stanford, CA.

McGee, M. (1979), 'Human spatial abilities: Psychometric studies and environmental, genetic, hormonal and neurological studies', *Psychological Bulletin* **86**(5), 889–918.

Meece, J., Parsons, J., Kaczala, C., Goff, S., & Futterman, R. (1982), 'Sex differences in math achievement: Toward a model of academic choice', *Psychological Bulletin* **91**(2), 324–48.

Moss, J. (1982), 'Towards equality: Progress by girls in mathematics in Australian secondary schools', Occasional Paper no. 16, Australian Council for Educational Research, Hawthorn, Vic.

Mura, R. (1987), 'Sex-related differences in expectations of success in undergraduate mathematics', *Journal for Research in Mathematics Education* **18**(1), 15–24.

Parker, L. (1984), Sex differences in mathematics: Trends in participation and achievement 1976–1983, Paper presented at the Annual Conference of the Australian Association for Research in Education, Perth, November.

Parker, L., & Offer, J. (1987), 'Girls, boys and lower secondary school achievement: The shifting scene 1972–1986', *Unicorn* **13**(3), 148–54.

Pattison, P., & Grieve, N. (1984), 'Do spatial skills contribute to sex differences in different types of mathematical problems?', *Journal of Educational Psychology* **76**(4), 678–89.

Pedersen, K., Elmore, P., & Bleyer, D. (1986), 'Parent attitudes and student career interests in junior high school', *Journal for Research in Mathematics Education* **17**(1), 49–59.

Peterson, P., & Fennema, E. (1985), 'Effective teaching, student engagement in classroom activities and sex-related differences in learning mathematics', *American Educational Research Journal* **22**(3), 309–35.

Quality of Education Review Committee (1985), *Quality of Education in Australia*, AGPS, Canberra.

Rallis, S., & Ahern, S. (1986), Math and science education in high schools: A question of equity, Paper presented at the 70th Annual Meeting of the American Educational Research Association, San Francisco April.

Reyes, L., & Stanic, G. (1988), 'Race, sex, socioeconomic status, and mathematics', *Journal for Research in Mathematics Education* **19**(1), 26–43.

Rheingold, H., & Cook, K. (1975), 'The contents of boys' and girls' rooms as an index of parents' behavior', *Child Development* **46**, 459–63.

Rosenthal, R., & Rubin, D. (1982), 'Further meta-analytic procedures for assessing cognitive gender differences', *Journal of Educational Psychology* **74**(5), 708–12.

Rosier, M. (1980), *Changes in Secondary School Mathematics in Australia: 1964–1978*, Australian Council for Educational Research, Hawthorn, Vic.

Russell, S. (1984), 'A captive audience?', *Mathematics in School* **13**(1), 31–4.

Sassen, G. (1980), 'Success anxiety in women: A constructivist interpretation of its source and its significance', *Harvard Educational Review* **50**(1), 13–24.

Science and Technology Education in Western Australia: A Matter of Concern (1987), A joint project of the Science Teachers' Association of Western Australia, Royal Australian Chemical Institute (WA Branch) and Australian Institute of Physics (WA Branch), Perth.

Schofield, H. (1982), 'Sex, grade level, and the relationship between mathematics attitude and achievement in children', *The Journal of Educational Research* **75**(5), 280–4.

Senk, S., & Usiskin, Z. (1983), 'Geometry proof writing: A new view of sex differences in mathematics ability', *American Journal of Education* **91**, 187-201.

Sherman, J. (1979), 'Predicting mathematics performance in high school girls and boys', *Journal of Educational Psychology* **71**(2), 242-9.

Shuard, H. (1982), 'Differences in mathematical performance between girls and boys', in W. Cockcroft, *Mathematics Counts*, HMSO, London.

Shuard, H. (1986), 'The relative attainment of girls and boys in mathematics in the primary years', in L. Burton (ed.), *Girls into Maths can go*, Holt, Rinehart & Winston, London.

Slack, D., & Porter, W. (1980), 'Training, validity, and the issue of aptitude: A reply to Jackson', *Harvard Educational Review* **50**(3), 392-401.

Smith, I. M. (1964), *Spatial Ability: Its Education and Social Significance*, Robert P. Knapp, San Diego.

Smith, S., & Walker, W. (1988), 'Sex differences on New York State Regents Examinations: Support for the differential course-taking hypothesis', *Journal for Research in Mathematics Education* **19**(1), 81-5.

Soberg, S., & Imsen, G. (1988), 'Gender and science education I', in P. Fensham (ed.), *Developments and Dilemmas in Science Education*, Falmer Press, Lewes, Sussex.

Stage, E., Kreinberg, N., Eucles, J., & Becker, J. (1985), 'Increasing the participation and achievement of girls and women in mathematics, science and engineeering', in S. Klein (ed.), *Handbook for Achieving Sex Equity through Education*, Johns Hopkins University Press, Baltimore, MD.

Stanworth, M. (1981), *Gender and Schooling: A Study in Sexual Divisions in the Classroom*, Hutchinson, London.

Taylor, H. (1986), 'Girls and boys and Lego', *Mathematics Teaching* **115**(June), 2-5.

Tittle, H. (1986), 'Gender research and education', *American Psychologist* **41**(10), 1161-8.

Thomas, B., & Costello, J. (1988), 'Identifying attitudes to mathematics', *Mathematics Teaching* **122**(March), 62-4.

Walden, R., & Walkerdine, V. (1982), 'Girls and mathematics: The early years', Bedfordway Papers no. 8, University of London Institute of Education, London.

Walden, R., & Walkerdine, V. (1986), 'Characteristics, views and relationships in the classroom', in L. Burton (ed.), *Girls into Maths can go*, Holt, Rinehart & Winston, London.

Walkerdine, V. (1983), 'Its only natural: Rethinking child-centred pedagogy', in A. Wolpe & J. Donald (eds), *Is there Anyone here from Education?*, Pluto Press, London.

Wattanawaha, N., & Clements, M. (1982), 'Qualitative aspects of sex-related differences in performance on pencil-and-paper spatial questions, grades 7-9', *Journal of Educational Psychology* **74**(6), 878-87.

Webber, V. (1987/8), 'Maths as a subversive activity', *Education Links* (32), 6-9.

Weiner, N., & Robinson, S. (1986), 'Cognitive abilities, personality and gender differences in math achievement of gifted adolescents', *Gifted Child Quarterly* **30**(2), 83-7.

White, J. (1986), 'Writing on the wall: Beginning or end of a girl's career', *Women's Studies International Forum* **9**(5), 561-74.

Widdup, D. (1980), 'Review of research on sex differences in mathematics', *Improving Maths for Girls*, Report of a conference held at Raywood Inservice Centre, Adelaide, May/June, Education Research and Development Committee.

Widdup, D. (1981), 'Women and maths: A wasting of talent', *Education News* **17**(6), 22–5.

Willis, S. (1985), 'Relating and integrating: An attempt', in J. Owen, N. Johnson & R. Welsh (eds), *Primary Concerns*, Commonwealth Tertiary Education Commission, Melbourne.

Woodrow, D. (1984), 'Cultural impacts on children learning mathematics', *Mathematics in School* **13**(5), 5–7.

Yates, L. (1988), 'Does "all students" include girls?: Some reflections on recent educational policy, practice and theory', *Australian Educational Researcher* **15**(1), 41–57.

Yates, J., & Firkin, J. (1986), *Student Participation in Mathematics: Gender Differences in the Last Decade*, Victorian Curriculum and Assessment Board, Melbourne.

Annotated bibliography

Buerk, D. 'The voices of women making meaning in mathematics'. *Journal of Education*, vol. 167, no. 3, 1985, pp. 59-70.
In this article Buerk raises issues related to the common understanding of mathematics as absolute and not 'person made'. While this view is inconsistent with the practice of mathematics, it is widely held and may explain some mathematics avoidance. The author works with adult women.

Burton, L. (ed.). *Girls into Maths*. Cambridge University Press, Cambridge, 1986.
This book is designed to be used as a package for teachers as part of a professional development program. Much of the material is based on the situation in the United Kingdom but it is none the less an excellent resource and could readily be adapted for and by Australian teachers.

Burton, L. (ed.) *Girls into Maths can go*. Holt, Rinehart & Winston, London, 1986.
This is a collection of readings on girls' mathematics education. Many of the articles are reprinted from elsewhere but their collection provides an excellent resource. The articles range from those which provide statistical data (UK) to the philosophical and the sociological, and from the conceptual to the applied. All of the articles are likely to be valuable and some will be more suited to your situation than others. However, I draw your attention particularly to the following papers as complementing or extending ideas and arguments presented in this monograph:

L. Burton, Introduction, pp. 1-20 — an excellent overview of both the field and the book.
R. Walden & V. Walkerdine, 'Characteristics, views and relationships in the classroom', pp. 122-46 — one chapter of an excellent report on the dissonance for girls between primary and secondary schooling and the impact of this dissonance on girls' access to mathematics.
L. Burton & R. Townsend, 'Girl-friendly mathematics', pp. 187-95 — a brief report on a strategy for changing girls' perspectives of mathematics and its relevence to their lives.
S. Brown, 'The logic of problem generation: From morality and solving to de-posing and rebellion', pp. 196-222 (also available in *For the Learning of Mathematics*, vol. 4, no. 1, 1984, pp. 9-20 — an inspiring article on what mathematics education might be.

Z. Isaacson, 'Freedom and girls' education: A philosophical discussion with particular reference to mathematics', pp. 223–40 — an important discussion of what freedom means in the context of educational and occupational choices.

Buxton, L. *Do you Panic about Maths?*. Heinemann Educational Books, London, 1981. The subtitle of this book is *Coping with Maths Anxiety*. It is very readable and very interesting from both a practical and a theoretical point of view.

Clements, M. 'Sex differences in mathematical performance: An historical perspective'. *Educational Studies in Mathematics*, vol. 10, 1979, pp. 305–22.

In this article Clements traces the history of girls' under-representation in mathematics in the United Kingdom. Of clear relevence to Australian education, it nicely places the present situation in its historical context.

Education Links, no. 32, 1987/8.

This whole issue on 'Demystifying maths and science' is likely to be of interest to teachers, but I draw your attention particularly to the following three articles:

V. Webber, 'Maths as a subversive activity', pp. 6–9 — raises issues regarding what a truly empowering mathematics curriculum might look like and reports on experiences the author has had with technical and further education classes in which she and her students have used mathematics to analyse aspects of society.

G. Joseph, 'Eurocentric mathematics', pp. 18–19 — provides extracts from G. Joseph, 'Foundations of eurocentrism in mathematics', *Race and Class*, vol. 27, no. 3, 1987, pp. 1–12. This is a fascinating article and I recommend that you make an effort to obtain it.

M. Barnes, 'The power of calculus', pp. 25–7 — suggests an alternative approach to the teaching of calculus. The author's ideas are applicable across a range of aspects of mathematics.

Fennema, E. (ed.). 'Explaining sex-related diferences in mathematics: Theoretical models'. *Educational Studies in Mathematics*, vol. 16, no. 3, 1985, pp. 303–20.

This is a collection of four short articles which present theoretical models on gender differences in mathematics:

G. Leder, 'Sex-related differences in mathematics: An overview', pp. 304–9.

E. Fennema & P. Peterson, 'Autonomous learning behavior: A possible explanation of sex-related differences in mathematics', pp. 309–11.

J. Eccles, 'Models of students' mathematics enrollment decisions', pp. 311–14.

D. Maines, 'Preliminary notes on a theory of informal barriers for women in mathematics', pp. 314–20.

For Australian data on participation rates see one of the following:

Dekkers, J., De Laeter, J., & Malone, J. *Upper Secondary School Science and Mathematics Enrolment Patterns in Australia, 1970–1985*. Curtin University of Technology, Bentley, WA, 1986.

Improving the Science, Mathematics and Technology Education Base in Australia. Report of the Working Group on Education for Science and Technology to the Victorian Government, Melbourne, 1987 (of particular interest to Victorian teachers).

Jones, W. 'Entry to engineering, technology, and science in higher education', Occasional Paper no. 20, Australian Council for Educational Research, Hawthorn, Vic., 1988.

Science and Technology Education in Western Australia: A Matter of Concern, A joint project of the Science Teachers' Association of Western Australia, Royal Australian Chemical Institute (WA Branch) and Australian Institute of Physics (WA Branch), Perth, 1987.